In Search of
Simulacra

Modeling of a Self Learning Android

In Search of Simulacra
Modeling a Self Learning Android

by Ted Di Salvo

123456789 HDHD 987654321

ISBN: 978-1-5136-5307-5

See all references at back of book

Acknowledgments

The author wishes to thank several people for their help in reviewing this book. My family Marian, Christine and Thea for their encouragement and my friend and former colleague Bill L'Hotta from the time we worked together at Bell Labs.

Table of Contents

Chapter 1 ... Can a Computer Think ?....................................... 1
Chapter 2 ... Can a Computer Synthesize Thought ?............... 9
Chapter 3 ... Thinking about thinking................................... 14
      a. Observation..15
      b. Analysis... 15
      c. Inference... 16
      d. Prediction..16
      e. Problem Solving... 17
      f. The problem-solving cycle in thinking..............18
Chapter 4 ... How could a Humanoid Android work ?........... 20
      a) Physical Motion Subsystems........................... 21
      b) Audio Subsystem.. 23
      c) Visual Subsystem.. 24
      d) Subsystems working together.........................25
      e) Recognition System.. 26
Chapter 5 ... What is Learning ?...27
      a. Simple Closed Loop System............................ 27
      b. Neural Network Systems................................. 30
      c. Expert Systems..33
      d. Search Techniques..37
          i. Binary Search Algorithm............................ 37
          ii. Linear Search Algorithm...........................39
          i. Heuristic Algorithm................................... 46
      e. Artificial Neural Networks................................ 50
          a. Group method of data handling............... 54
          b. Convolutional neural networks................. 54
          c. Long short-term memory..........................54
          d. Deep reservoir computing.........................55
          e. Deep belief networks................................55
          f. Large memory storage and retrieval neural
          networks... 56
          g. Stacked (de-noising) auto-encoders........... 56
          h. Deep stacking networks........................... 57
          i. Tensor deep stacking networks..................57
          j. Spike-and-slab RBMs................................. 57
          k. Compound hierarchical-deep models.........58
      f. Object Oriented System Architecture............... 58
      g. Goal Oriented Systems Architecture................. 61

Chapter 6 ... Building Blocks.. 66
   a. Self Learning Cell - the smallest unit........................67
   b. Self Learning Objects,.. 68
Chapter 7 ... Motion Objects...73
   a. Legs... 73
   b. Arms..76
   c. Other... 78
Chapter 8 ... Visual Objects.. 79
   a. Automatic image storage....................................80
   b. Object recognition.. 80
   c. Obstacle avoidance... 80
   d. Video image comparisons................................... 81
   e. Distance measurement.......................................81
   f. Common sensory functions.................................81
Chapter 9 ... Audio Objects... 83
   a. Automatic sound storage....................................83
   b. Loud noise detection... 83
Chapter 10 ... Odor, Balance and Tactile Sensor Objects....... 85
   a. Odor detection..85
   b. Tactile sensor...85
   c. Balance sensor... 86
Chapter 11 ... Memory Objects...87
Chapter 12 ... Object Manager...89
Chapter 13 ... Motivational Objects......................................90
Chapter 14 ... Goal Objects... 95
Chapter 15 ... Learning Objects..97
Chapter 16 ... Analysis Objects.. 98
   a. Curiosity begets learning................................... 98
   b. Formal Learning.. 100
   c. Decisions, Decisions, Decisions...........................101
   d. The data in the database.................................. 104
Chapter 17 ... Other Issues..105
Table of figures...107
References... 109

i. Introduction

A conceptual model's primary objective is to convey the fundamental principles and basic functionality of the system which it represents (the theory of operation). This document attempts to model a self learning android from top to bottom. Unlike a computer simulation of human behavior, this model seeks to define a simulation of the general purpose underlying processes (behavior) which humans normally follow. This underlying process attempts to mimic a human's learning ability from early childhood through to adult. In other words, this model does not explain how to write a program which can play checkers, rather it defines a program which allows the android to learn to play checkers.

This book attempts to explain, in plain language, the concepts necessary to describe the theory of operation of a self learning android. The only prerequisite is an interest in Artificial Intelligence. Please enjoy the journey.

Man has always had a certain fascination with robots. The ancient history of robotics from *slideshare.net* shows this.

History of Robotics

- **3500 BC -** Greek myths of Hephaestus and Pygmalion incorporate the idea of intelligent mechanisms. Something we would later call robots. (16)

- **2500 BC -** Egyptians invent the idea of thinking machines: citizens turn for advice to oracles, which are statues with priests hidden inside. (16)

- **1400 BC -** Babylonians develop a water clock named the "clepsydra." (16)

- This water clock is considered one of the first "robotic" devices in the history of man kind. The water is recycled through a kind of siphoning system. (16)

- **700 - 800 BC -** First symbolic mention of robots (automated) appears in Homer's Iliad - or simulacra as they will be called later. (16)

And now a bit more contemporary:

Artificial Intelligence - From Wikipedia, the free encyclopedia

> In computer science, **artificial intelligence (AI)**, sometimes called **machine intelligence**, is intelligence demonstrated by machines, in contrast to the **natural intelligence** displayed by humans and animals. Computer science defines **AI** research as the study of "intelligent agents": any device that perceives its environment and takes actions that maximize its chance of successfully achieving its goals. Colloquially, the term "artificial intelligence" is used to describe machines that mimic "cognitive" functions that humans associate with other human minds, such as "learning" and "problem solving".

> As machines become increasingly capable, tasks considered to require "intelligence" are often removed from the definition of AI, a phenomenon known as the **AI effect**. A quip in Larry Tesler's[1] theorem says **"Artificial Intelligence is whatever hasn't been done yet."** For instance, optical character recognition is frequently excluded from things considered to be AI, having become a routine technology. Modern machine capabilities generally classified as AI include successfully understanding human speech, competing at the highest level in strategic game systems (such as chess and Go), autonomously operating cars, intelligent routing in content delivery networks, and military simulations.

---

[1] Larry Tesler computer scientist who works in the field of human—computer interaction. Tesler has worked at Xerox PARC, Apple, Amazon, and Yahoo!.

The following are some common types of artificial intelligence taken from - Simplicable.com

Activity Recognition

Determining what humans or other entities such as robots are doing. For example, a car that can see its owner approaching with a heavy bag of groceries may decide to open an appropriate door automatically.

Affective computing

*is a type of artificial intelligence that seeks to understand and use emotion. It is the machine equivalent of emotional intelligence in humans. Affective computing has numerous applications related to improving computer-human interactions and interpreting human actions and behaviors. Artificial intelligence can potentially detect human emotions using cues such as facial expressions, tone of voice, word usage or other inputs such as the way someone is typing or their body temperature. As with any powerful tool, this has potential to improve quality of life or make it worse depending on how it is applied and governed. **Affective computing has broad privacy implications such as your toaster could theoretically record your emotions and make this information available to companies and/or governments.** Alternatively, emotions could be unrecorded and simply used to improve user interfaces such as a synthetic voice that can reflect a range of emotions in its tone.*

Artificial Creativity

*is any **AI** application which is in those areas that are viewed as creative, such as music, art and design. This also applies to human-like abilities such as: including humor in a conversation and interpreting human actions and behaviors. Artificial intelligence can potentially detect human emotions using cues such as facial expressions, tone of voice, word usage or other inputs such as the way someone is typing or their body temperature. As with any powerful tool, this has* *potential to improve quality of life or make it worse depending on how it is applied and governed. Affective computing has broad privacy implications as your toaster could theoretically record your emotions and make this information available to companies and/or governments. Alternatively, emotions could be unrecorded and simply used to improve user interfaces such as a synthetic voice that can reflect a range of emotions in its tone.*

Artificial Intelligence from the Author's perspective:

Three things are required in order to have artificial intelligence closely mimic human intelligence:

> ➢ The ability to learn from experience

> ➢ It would need to have all the sensory input and output that a human has

> ➢ It would need to have all the physical ability and restrictions that a human has

If Artificial Intelligence were to try to mimic a cat, it would need the same three things:

> ➢ The ability to learn from experience

> ➢ It would need to have all the sensory input and output that a cat has

> ➢ It would need to have all the physical ability and restrictions that a cat has

The physical characteristics and sensory abilities of any intelligent entity is an integral part of its intelligence and therefore must be present for an Artificial Intelligence to mimic it.

Chapter 1 ... Can a Computer Think ?

NO, say most experts. Some quotes from experts are:

"There's more to intelligence than processing speed. While a supercomputer like the Sequoia can analyze problems and reach a solution faster than humans, it can't adapt and learn the way humans can. Our brains are capable of analyzing new and unfamiliar situations in a way that computers can't. We can draw upon our past experiences and make inferences about the new situation. We can experiment with different approaches until we find the best way to move forward. Computers aren't capable of doing that -- you have to tell a computer what to do."

Well, if a computer only does what it is programmed to do, then direct it to think. Problem solved.

Coined by computing pioneer Alan Turing in 1950, the Turing test was designed to be a rudimentary way of determining whether or not a computer counts as "intelligent". The test, as Turing designed it, is carried out as a sort of imitation game. On one side of a computer screen sits a human judge, whose job is to chat to some mysterious interlocutors on the other side. Most of those interlocutors will be humans; one will be a chatbot, created for the sole purpose of tricking the judge into thinking that it is the real human.

Some computer scientists are attempting to design computers that can mimic human thought -- a tricky situation considering we still don't have a complete understanding of how we think. Other computer scientists prefer to design systems that don't use the brain as a model. Futurists like Dr. Ray Kurzweil predict *that it's just a matter of time before we develop a computer system capable of being self-aware. After that, we may see computers capable of **recursive self-improvement**. That means computers will be able to analyze their own capabilities and make adjustments to improve performance.*

The following posted by John Spacey, March 30, 2016 updated on January 08, 2017

***Recursive self-improvement** describes software that writes its own code in repeated cycles of improvement. It is associated with artificial intelligence as self-improving software has potential to develop superintelligence. Traditional artificial intelligence is coded by humans. Each AI develops its own intelligence represented with data and formulas such as non-linear transformations. Traditional AI is able to learn but can't fundamentally change its own design to become something new. A recursive self-improving program could theoretically develop superintelligence and aspects of consciousness such as intentionality. This is typically thought to represent an existential risk because such intelligence may develop goals that conflict with the interests of humans. Being more intelligent than humans, a superintelligence could be a significant threat to human quality of life and survival.*

Before worrying about the **Terminator model 3.0**[2], let's start with the basics: Can a computer think?
First of all, it would depend on the definition of thinking. Thinking does not have to be human thinking, whatever that is. Does an animal think? Yes, well maybe not the way humans do. But an animal can make decisions, show emotions, be creative in looking for food, protection, etc.. Isn't that thinking ? If it isn't thinking, it seems close enough.

Do newborn human infants think? Yes, they most certainly do, as shown in the figure 1 chart. From casually observing a

---

[2] Terminator is a fictional character from a movie by the same name

2

newborn, there doesn't seem to be much thinking going on initially. However, thinking does become evident in short order. A newborn baby figures out how to get fed, get changed, picked up etc.. very early on. Most of them outwit their parents within a month or so.

The following is from How it Works Magazine 5/2019 based on a US Government study
Stages of brain development in an infant

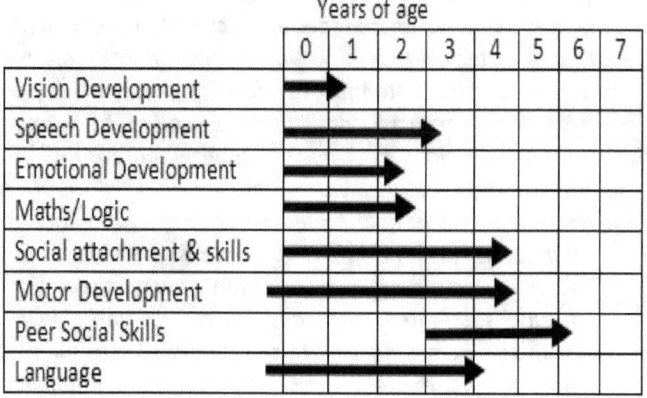

	Years of age							
	0	1	2	3	4	5	6	7
Vision Development	→							
Speech Development			→					
Emotional Development			→					
Maths/Logic			→					
Social attachment & skills					→			
Motor Development					→			
Peer Social Skills						→		
Language			→					

Figure 1 - Human Brain Early Development

* Interestingly enough, some thinking is going on before the infant is born as shown in this chart.

Back to animals, most newborn animals can survive on their own within minutes, hours, days or months. They are endowed with what humans refer to as instincts at birth. A newborn horse, for example, literally lands on their feet and can walk, albeit a bit rocky in moments. Baby chicks, chirp to get fed from their mother. Baby wood ducks jump out of their nest with-in weeks of birth and crash land in a pond and swim. There are certainly many more examples of newborn animal's **instincts[3].**

---

[3]For more examples of animal instincts check the following web site: **https://www.reference.com/science/examples-animal-instincts-3a af16173be360ca**

Humans and animals have a distinct advantage over a machine in that they have built in motivations. Hunger is the probably the biggest motivation of all, right from the start with humans and animals. Motivation may be the "**mother of thinking**". Some other motivators would be discomfort, danger and wanting to please the parent or teacher or conversely not wanting to displease other people. Animals learn to hunt for food by following the example of their parent. Human babies learn to please their parents by making them smile.

Machine thinking would certainly be different from biological thinking, but it may be thinking none the less. If the machine can come up with unique solutions to problems, which it has not been programmed to do, then it may be "thinking".

Of course we have been warned about unbridled, thinking robots, by science fiction books and movies over and over again. So perhaps we shouldn't strive for **very clever** humanoids anyway.

Animals have instinctive behavior and humans have brain power or do we?

Excerpts below taken from internet article by By Peter Aldhous 10 Feb 2015

Human brain size compared to animals

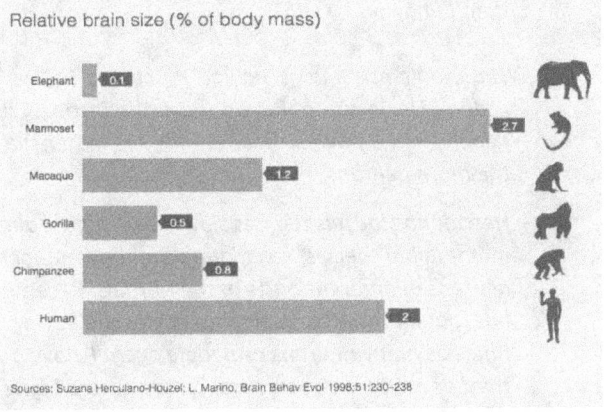

Figure 2 - Human Brain vs Animal Brain

Absolute brain size clearly is not what really matters, otherwise people would be cognitive pygmies compared to whales and elephants. But brain size relative to body size does not seem to be a particularly informative measure either. Marmosets which are a diminutive monkey that are not thought to be among the brightest of primates, have brains that tip the scales at around 2.7 percent of their body mass, easily beating humans 2.0 percent.

By the 1970s, the answer seemed to have arrived in the form of a measure called **encephalisation quotient***, or EQ. Its inventor, Harry Jerison at the University of California, Los Angeles, realized that the relationship between brain and body size is not linear - smaller animals usually have brains that are relatively larger. So his EQ formula considers how an animals brain compares to the predicted size, given the scaling curve for the group to which it belongs.*

At last, there was a measure that elevated people to the pinnacle. Our EQ is an impressive 7.0 - meaning that our brains are seven times larger than expected for a mammal of our size. But the numbers for our primate relatives do not make much sense. Rhesus monkeys, by this measure, should be smarter than gorillas - which does not seem to be the case, based on their behavior.

The problem, explains Suzana Herculano-Houzel of the Federal University of Rio de Janeiro in Brazil, is that EQ depends on a flawed assumption: The larger the brain, the more neurons.

Herculano-Houzels team is slowly working through mammalian species, calculating the total number of neurons in different parts of their brains. Recently, she has found that African elephants, despite having three times as many neurons as people overall, have only a third as many in the cerebral cortex, where **higher** cognitive functions reside.

Species	Brain size (g)	Relative brain size (% of body mass)	Encephalisation quotient	Brain neurons (billions)	Cerebral cortex neurons (billions)
Human	1330	2	7	86	16.3
Chimp	390	0.8	2.5	22	6
Gorilla	500	0.5	1.6	33	9.1
Rhesus monkey	93	1.2	2.1	6	1.7
Marmoset	8	2.7	1.7	0.634	0.245
Elephant	4148	0.1	1.2	251	5.6

Sources: Suzana Herculano-Houzel; Marino, L. Brain Behav Evol 1998;51;230-238

Figure 3 - Comparison of Neuron in Cerebral Cortex

Human brains have the highest number of neurons in the cerebral cortex. Higher, than any other member of the animal kingdom. So maybe that's why we seem to have less in the way of instinctive behavior at our early stages. Because we can more than compensate later on in our development.

How Many Tasks Can Our Brains Process At The Same Time?
Neuroscientist Harris Georgiou from the National Kapodistrian University of Athens in GreeceGeorge Dvorsky 11/10/14 12:30 pm Filed to: NEUROSCIENCE

The human brain has been described as a massively parallel computing machine. But just how powerful is it? A recent brain scan analysis is offering some unexpected results.

Although the analysis is complex, the outcome is simple to state. Georgiou says that independent component analysis reveals that about 50 independent processes are at work in human brains performing the complex visuomotor[4] tasks. However, the brain uses fewer processes when carrying out simple tasks, like visual recognition.

That's a fascinating result that has important implications for the way computer scientists should design chips intended to mimic human performance. It implies that parallelism in the brain does not occur on

---

[4] Visuomotor - of or relating to vision and muscular movement

the level of individual neurons but on a much higher structural and functional level, and that there are about 50 of these.

*Georgiou points out that a typical voxel[5] corresponds to roughly three million neurons, each with several thousand connections with its neighbors. However, the current state-of-the-art **neuromorphic chips** contain a million artificial neurons each with only 256 connections. What is clear from this work is that the parallelism that Georgiou has measured occurs on a much larger scale than this.*

So an artificial equivalent of a brain-like cognitive structure may not require a massively parallel architecture at the level of single neurons. Instead, and as noted by Georgiou, one could be built using "a properly designed set of limited processes that run in parallel on a much lower scale."

What Neuro-scientist Harris Georgiou's research is saying is that the brain actually works in a compartmentalized manner. That is small segments of the brain's activity, take care of different problems. for example, a visual processor, a speech processor, a memory processor, learning processor and on and on. This is good news for modeling the brain, making the models task based. More about this later.

Neuromorphic Chips: Microchips that Imitate the Brain
By UNIVERSITY OF ZURICH JULY 23, 2013

Study demonstrates how complex cognitive abilities can be incorporated into electronic systems made with neuromorphic chips.

Novel microchips imitate the brains information processing in real time. Neuroinformatics researchers from the University of Zurich and ETH Zurich together with colleagues from the EU and US demonstrate how complex cognitive abilities can be incorporated into electronic systems made with so-called neuromorphic

---

[5] Voxel - abbreviation for volume element, the three-dimensional version of a pixel.

chips: They show how to assemble and configure these electronic systems to function in a way similar to an actual brain.

Neuromorphic engineering From Wikipedia, the free encyclopedia

__Neuromorphic engineering__, also known as __neuromorphic computing__ is a concept developed by Carver Mead in the late 1980s, describing the use of very-large-scale integration (VLSI) systems containing electronic analog circuits to mimic neuro-biological architectures present in the nervous system. In recent times, the term neuromorphic has been used to describe analog, digital, mixed-mode analog/digital VLSI, and software systems that implement models of neural systems (for perception, motor control, or multisensory integration). The implementation of neuromorphic computing on the hardware level can be realized by oxide-based memristors[6], spintronic[7] memories, threshold switches, and transistors.

---

[6] A memristor is a hypothetical non-linear passive two-terminal electrical component relating electric charge and magnetic flux linkage. It was envisioned, and its name coined, in 1971 by circuit theorist Leon Chua

[7] Spintronics, also known as spin electronics, is the study of the intrinsic spin of the electron and its associated magnetic moment, in addition to its fundamental electronic charge, in solid-state devices

Chapter 2 ... Can a Computer Synthesize Thought ?

Again the answer is **not really** by most experts.

However, consider the following excerpts taken from the internet, which tend to support the notion of synthesized human thought.

> Humans are very good at recognizing patterns. While we're making progress in machine pattern recognition, it's mostly on a superficial level. For example, some digital cameras can recognize specific faces and automatically tag photos of those people as you take pictures. But humans can recognize complex patterns and adapt to them -- computers still have trouble doing that.

> One of the features computers would need to be more intelligent than humans is the ability to draw conclusions from observations. In a study published in 2009, computer engineers at Cornell University designed a program that could do this on a limited scale. The program gave the computer a basic set of tools it could use to observe and analyze the movements of a pendulum. Using this foundation, the software was able to extrapolate basic laws of physics from the pendulum's motions. It took about a day for the computer to arrive at the same conclusions it took humans thousands of years to grasp [source: Steele].

The following are excerpts taken from Can Computers Think? by Frank Buytendijk Dec. 27, 2012

> IBM's computer, Deep Blue, did defeat chess champion Garry Kasparov. The programming of Deep Blue was largely one of brute force, calculating an unbelievable

200 million positions per second and "thinking through" six to eight moves in advance. Additionally, Deep Blue contained a huge library of 700,000 chess games. The developments didn't stop with Deep Blue's victory. Today's chess software running on standard PCs may not calculate as many moves per second, but contain much smarter algorithms.

Moreover, IBM has successfully moved beyond chess and has succeeded in a much, much harder domain. In 2011, the IBM Watson computer won the American game show *Jeopardy!*, beating the two best contestants the show ever had. At the core of Watson was an engine to parse language, including trying to understand clever word play and slang, on which many of the show's questions hinge. The computer's programming was a combination of many different styles of algorithms trying to interpret the questions, in combination with four terabytes of semantically structured information – an infinitely more complex task than playing chess. Still, Watson is far from passing the Turing test. It may interpret language better than any other computer, but it is focused on providing answers instead of full conversation.

The following are excerpts from
www.computersciencedegreehub.com

The Difference Between Brains and Computers

The human brain is a highly complex unit. It doubles and folds back upon itself to provide room for the neurons that process all the data that we absorb daily. According to **2Machines.com**, there are a million, million neurons in the adult human brain, each with a thousand connections. The neurons process data in a parallel method, making it possible to analyze and make decisions quickly. To replicate that complexity, computers would have to have 8 million Gigabytes of RAM. Humans are able to take in complex data like color, movement, sound and size and process them simultaneously to arrive at an identification or to *make a response. The computer processes each area individually, assimilates the data to identify the subject, then must still rely on pre-programmed data to respond. At the current rate of technological advancement, computers should have eight million Gigabytes by 2029, so can we assume that they will be able to* **think** *at that point?*

The following is an excerpt from an internet article: Can Computers Think Creatively? By Mark McGuiness

Consider the findings of an experiment <u>reported in the New York Times</u>, *in which humans were pitted against computers to see who could come up with the best ideas for advertisements.*

The humans were non-advertising professionals, given a brief and asked to come up with creative ideas for adverts. The computers were programmed with an algorithm for devising advertising ideas and given the same briefs.

Here's a sample of the results:

- **COMPUTER IDEA:** *An Apple computer offers flowers (for advertising Apple Computers' friendliness).*
- **HUMAN IDEA:** *An Apple computer placed next to a PC with the claim: This is the friendliest computer.*
- **COMPUTER IDEA:** *Two Jeeps communicating in sign language (for advertising a silent car engine).*
- **HUMAN IDEA:** *A car driving alone in the country.*
- **COMPUTER IDEA:** *A domed mosque with tennis ball texture (for World Cup Tennis tournament in Jerusalem).*
- **HUMAN IDEA:** *A picture of ancient walls of Jerusalem with a tennis poster on them.*

I think most of us would call that 3 to 0 for the computers.

The research panel agreed: they judged the computer ads to be consistently more original and creative than those devised by the human group.
What is going on here?
Does this mean the beginning of the end for human creative superiority? Not necessarily.

The obvious answer to **can computers synthesize human thought** seems to be **sort of yes**, computers can synthesize human thought. But what has been demonstrated has always been limited in scope. Focused in relatively narrow areas. What is missing is the overview of thought. The motivation of the thought or the reason to think about something. That is, should the machine think about one thing versus another.

If a computer had an immense amount of information about a myriad of subjects, it still could not think, unless it had the motivation to think about one thing or another. The computer needs to have goals to achieve, I.e. motivation to get anything done.

That sounds a bit like some people when they have nothing to do, they daydream, vegetate or watch TV, sometimes mindlessly. Then, someone comes along and says why have you not taken out the garbage or cleaned your room or something else. That is called motivation. The motivation to get the brain muscle in gear and start thinking.

The interesting thing about human thought is, conceivably, it could all be related to one or two built in **motivations:** hunger and pleasing (or the fear of not pleasing) a parent or being accepted by others. Those two motivators could be the underlying basis of everything that comes later in life, like playing sports, finding a mate, trying to get a better job, bigger house, going on vacations, etc..

The *goals* or the *focus* of these computers has, to date been programmed in by the **humans**. Solve one problem or the other, accomplish one task or another and on and on.

It would seem what would give the computer the ability to synthesize human thought is having the computer pick it's own topic or task to focus on. So, should the computer pick it's own tasks, maybe, maybe not.

Certainly a safer, more controlled way to go, would be if the tasks came from a large batch of previously selected potential tasks given by humans. Or some other safe guarded methodology to keep the computer's simulated thoughts from going too far afield. More about this later.

Chapter 3 ... Thinking about thinking

Merriam-Webster Dictionary says:
Thinking - *noun*
Definition of thinking: the action of using one's mind to produce thoughts

Thought - *noun*
Definition of thought:
a : an individual act or product of thinking
b : a developed intention or plan

So, there you have it, **thinking** is having a **thought** and a
thought is the result of **thinking.**

Based on that circular logic statement there is nowhere to go from here. Therefore, let's step back and break things down a bit more. There are different types of thought, like positive or negative or other sub types of thinking. Perhaps by taking a smaller bite of the **thinking** pie, we can find something more manageable.

Some types of thinking (taken from excerpts found on the internet)

I.   **Critical thinking** *is the mental process of objectively analyzing a situation by gathering information from all possible sources, and then evaluating both the tangible and intangible aspects, as well as the implications of any course of action.*
II.   **Implementation thinking** *is the ability to organize ideas and plans in a way that they will be effectively carried out.*
III. **Conceptual thinking** *consists of the ability to find connections or patterns between abstract ideas and then piece them together to form a complete picture.*

IV.   *Innovative thinking* involves generating new ideas or new ways of approaching things to create possibilities and opportunities.

V.   *Intuitive thinking* is the ability to take what you may sense or perceive to be true and, without knowledge or evidence, appropriately factor it in to the final decision.

If computer synthesized thinking can be achieved, perhaps it should be, at least initially, limited to Critical thinking.

Various excerpts from the internet on Critical thinking for example, can be defined as follows:

a.   **Observation**

A key component of critical thinking skill involves observation -- you use this to gather information about a process, for example. There are basically two types of observation, direct and participant. Direct observers try not to engage with a process, while participant observers may interact. This could be the difference between a teacher sitting in a classroom observing or watching behind a two-way mirror, for example. Each method is an attempt to obtain information in the most objective way possible, however some scholars suggests that people can never completely divorce themselves from the processes that formulate them -- their own experiences, and cultural collective experiences such as language or even religion essentially negate the ability to remain objective entirely, supplant another set of value systems for their own.

b.   **Analysis**

Analysis is another component of critical thinking. It is a way to evaluate what it is that you observe, for example. Analyzing information entails review and organization. You can break down these components a bit further, as well. People often compare and contrast what it is they have discovered, then decide what to do with the information they obtain -- afterward, they may begin to integrate components together synthesizing information in a new way.

c. Inference

Inference is the manner in which people make informed conclusions. It is a bit different than assumption -- the method whereby people make conclusions based on what they assume to be true rather than on what they learn. This is not to suggest that inferences are objective or entirely fact-based as what people often interpret as inference sometimes varies cross-culturally, for example. Both the context and content of inferences may change over the span of generations, as well.

d. Prediction

Critical thinkers may eventually come to a point when they have to apply what they know either in thought or deed. This type of critical thinking skill is called prediction -- the method of applying your inferences to essentially guess what will happen. Different professional fields apply critical thinking models that involve prediction virtually all the time: a statistician uses quantitative information to generate projections for a business, for example.

Excerpts from Encyclopedia Britannica **Types of thinking**

*Philosophers and psychologists alike have long realized that thinking is not of a **single piece**. There are many different kinds of thinking, and there are various means of categorizing them into a **taxonomy** of thinking skills, but there is no single universally accepted taxonomy. One common approach divides the types of thinking into problem solving and reasoning, but other kinds of thinking, such as judgment and decision making, have been suggested as well.*

e. Problem Solving

Problem solving is a systematic search through a range of possible actions in order to reach a predefined goal. It involves two main types of thinking: divergent, in which one tries to generate a diverse assortment of possible alternative solutions to a problem, and convergent, in which one tries to narrow down multiple possibilities to find a single, best answer to a problem. Multiple-choice tests, for example, tend to involve convergent thinking, whereas essay tests typically engage divergent thinking.

Elsewhere on the Internet
*   **Convergent thinking** is the process of finding a single best solution to a problem that you are trying to solve. Many tests that are used in schools, such as multiple choice tests, spelling tests, math quizzes, and standardized tests, are measures of convergent thinking. Traditional intelligence tests, such as the Stanford-Binet, also measure convergent testing.*

*   **Divergent thinking** means thinking that starts from a common point and moves outward into a variety of perspectives*

*   Whenever we use **divergent thinking**, we search for options instead of just choosing among predetermined options. **Convergent thinking** relies heavily on logic and less on creativity, while divergent thinking emphasizes*

creativity. Divergent thinking works best in problems that are open-ended and allow for creativity.

Again excerpts from Encyclopedia Britannica **Types of thinking**

f.   The problem-solving cycle in thinking

Many researchers regard the thinking that is done in problem solving as cyclical, in the sense that the output of one set of processes - the solution to a problem- often serves as the input of another- a new problem to be solved. The American psychologist Robert J. Sternberg identified seven steps in problem solving, each of which may be illustrated in the simple example of choosing a restaurant:

a. Problem identification. In this step, the individual recognizes the existence of a problem to be solved: he recognizes that he is hungry, that it is dinnertime, and hence that he will need to take some sort of action.

b. Problem definition. In this step, the individual determines the nature of the problem that confronts him. He may define the problem as that of preparing food, of finding a friend to prepare food, of ordering food to be delivered, or of choosing a restaurant.

c. Resource allocation. Having defined the problem as that of choosing a restaurant, the individual determines the kind and extent of resources to devote to the choice. He may consider how much time to spend in choosing a restaurant, whether to seek suggestions from friends, and whether to consult a restaurant guide.

d. Problem representation. In this step, the individual mentally organizes the information needed to solve the problem. He may decide that he wants a restaurant that meets certain criteria, such as close proximity, reasonable price, a certain cuisine, and good service.

e. *Strategy construction. Having decided what criteria to use, the individual must now decide how to combine or prioritize them. If his funds are limited, he might decide that reasonable price is a more important criterion than close proximity, a certain cuisine, or good service.*

f. *Monitoring. In this step, the individual assesses whether the problem solving is proceeding according to his intentions. If the possible solutions produced by his criteria do not appeal to him, he may decide that the criteria or their relative importance needs to be changed.*

g. *Evaluation. In this step, the individual evaluates whether the problem solving was successful. Having chosen a restaurant, he may decide after eating whether the meal was acceptable.*

The above steps **a** thru **g** are starting to sound logical, like the scientific method, maybe programmable. Reminiscent of Object Oriented computer programming languages. More about this later.

Chapter 4 … How could a Humanoid Android work ?

Bob, our android, looks like a humanoid, two arms, two legs, eyes, ears and nose. Bob started off his existence not knowing anything like most newborns. Of course, unlike a newborn, Bob was full grown in size at the time. Newborns have certain involuntarily mechanisms working from the get go, like heartbeat and breathing. Bob doesn't need either one of those.

Newborns feel certain things like hunger and/or discomfort which would provide motivation or goals for the newborn. Bob won't feel hunger since he doesn't eat, but he does feel pain since he has sensors on his skin. And like a newborn, Bob should seek to please his trainer/guardian and thus providing Bob's motivation or goal. This goal to please or not displease would be pre-programmed into the Android. More on this later.

In order to make Bob "grow up" he needs a lot of training especially in the beginning. He needs to learn to walk, talk, play catch, etc.. To visualize how this is possible let's back up a bit and delve into Bob's basics.

Bob has ostensibly all the mechanics an Android would need to walk, talk, listen, see, do tasks. That is Bob has the equivalent of bones and muscles. Also, eyes that can see, nose to smell and ears that can listen. Bob further has sensors in his skin to detect pressure, temperature and pain.

Bob's physiologic break down, if you will, is he has numerous subsystems controlling the movement of his skeletal "bones". Each subsystem does a portion of a task, I.e. the foot subsystem provides some balance and adjusts to the contour of the ground beneath his feet. The attached leg subsystem provides further balance and adjust the load of the upper body as perhaps Bob is walking, running, climbing, standing still in a wind storm, etc.

There would of course be many more subsystems to allow Bob to function, like an individual toe subsystem and individual finger subsystem. A foot subsystem, a hand subsystem. A forearm subsystem, a calf subsystem. A thigh subsystem, an upper arm subsystem. And on and on all the way up to an ear subsystem, and eyebrow subsystem, a smile subsystem, etc., etc., etc..

Each subsystem does its localized job in response to a higher level directive. For example, walking would require the overall directive to "go to the window". The toes would report to the foot, the foot to the ankle, the ankle to the calf, the calf to the thigh, the thigh to the hip, etc., etc. All the way up to the vision (for obstacle avoidance). And all of these subsystems would be working to accomplish the goal of walking to the window.

a)   Physical Motion Subsystems

Looking closely at a single subsystem, it needs to "learn" how to perform and "remember" the results to use again in a similar situation. That would be accomplished by trial and error. If you try to grab a baseball you must contour the fingers to encircle the ball and apply enough pressure to hold the ball from slipping out of your grasp. This is learned by trail and error and so too the android will learn it that way as well. Much more about this later.

In the baseball example, each section of the individual finger would have a "controller" moving the "muscle" to tighten finger around the ball. The sensors in the skin would provide feedback to the subsystem as to when the muscle has moved enough to grasp the ball so it won't slip. The hand subsystem would also be involved in getting each of the finger subsystems to move at the same time to accomplish the overall task. Perhaps the wrist subsystem and or arm subsystem would need some involvement as well. At a higher level, which will be discussed later, the vision subsystem would also come into play. Obviously the number of subsystems involved in the hand holding something can be sizeable. The number of different things the hand could hold could be large as well.

As with the human child, the android would have to learn by trial and error a number of variations in order to successfully hold numerous items in its hand. Further, the system should remember each of the parameters for the different situations.

Now consider the number of subsystems required in the entire Android's body to mimic the human body and the number becomes rather large. In fact the human body has about 206 bones and 600 muscles. Therefore, the android would need somewhere in the order of 600 "muscles" to move the 206 structural "bones" in order to closely mimic the human body movement capability.

Each of these subsystems would go thru fairly extensive training or learning over time. Consider what happens to a human infant. They have to learn just about everything. They do have some built in traits which provide the motivation for learning. A newborn knows to cry in order to get fed or when they are uncomfortable. Early on they learn to kick their legs, throw their arms around. In doing so, they bump into things which may give them some pain so they learn to kick better and throw their arms around better. They stick their fingers and other things, which they learned to pick up, in their mouth to simulate eating. In other words, all those subsystems are constantly being trained from the very beginning. And they are being refined in their efficiency as time goes on.

The infant's vision is also being trained along with the hand/arm picking up things then sticking it in their mouth. That *is* hand eye coordination. All of this seems to be mostly trial and error, since the infant doesn't know the hand/arm is more efficient at picking up food and putting it in it's mouth than the foot/leg is. Although, I have seen a toe in the mouth on occasion.

Just to point out that some "newborns" in the animal kingdom are a bit more advanced in their initial abilities. A number of our four legged friends can walk from birth. They may initially be a bit wobbly, but have the basics and improve in relatively short order. Animals are born with instinctive characteristic, certainly more than the human newborns. A lot of the newborn animals can and do survive on their own, whereas a human infant could not survive without help. So some characteristics could and should be initially given to Bob so we don't have a full size humanoid flailing it's arms and legs all over the place. He would still need to refine his initial skills.

Bob would go thru the same kind of learning. With all the capabilities in place, that is the "muscles" and the "sensors", a "motivation" is all that is needed for the "learning" to begin. The motivation can come in many forms, a fly landing on Bob causing him some discomfort. Perhaps Bob attempting to please or mimic his trainer smiling at him and trying to smile back, or saying "hello Bob" in the morning.

Regardless of the stimuli, it will take time, as with the infant, for Bob to learn.

b)   Audio Subsystem

Just as an infant is taught to say the words, so will Bob learn to speak. By hearing the sound of a word, seeing the mouth of the teacher move, Bob will learn to speak via the hearing/speech/visual subsystems working together. Like an infant, Bob will probably start with making noises rather than intelligible words.

By the time a child is ready for school (around 5 years of training at home) they already have learned a lot, which forms a foundation for future learning. They know how to answer simple questions and ask for things. Additionally, children will recognize the voices, moods of the speaker (happy, sad or not pleased) and sounds (dogs barking, birds chirping, etc.). This is building up a memory of sounds. Of course young children go on to school to learn ABC's and other verbal skills. An Android probably would be home schooled. Perhaps Bob will take on-line courses.

Learning speech will require a large amount of memory since there is a vast amount of information Bob should eventually know. Probably Bob will additionally tap into the internet to access a lot of factual information.

c)   Visual Subsystem

Two eyes on the front plane of the head allow for, with training, depth perception. Since the eyes can move side to side or up and down independently of each other there is a subsystem for that. Also, each eye can focus for close up of distance, each would have a subsystem to control that. Mostly the eyes would be coordinating their movements and focus. A higher level subsystem would be requesting the eyes to focus on some object and collect images.

Depth perception is extremely useful in picking up objects and putting it in your mouth. In fact, it is vital in all instances, like catching a baseball, eating dinner or driving a car. Further, the visual system allows Bob and the child to recognize faces, objects and other things. This will end up requiring a lot of memory storage in both the child and Bob. To put some of this in perspective, an average human adult has approximately 100 billion neurons in their brain. Note, there are different estimates of the number of neurons from 86 billion to 120 billion. Whatever the actual number is, it's a lot of storage capacity.

d)   Subsystems working together

Almost all movement has hand-eye coordination at it's core. Catch a baseball, walk to the door, open a window, skip rope, and on and on. Catching a baseball for example, would have lots of subsystem inter-dependency. The young baseball player would see the ball coming towards them, have to judge it's speed and trajectory. They must move their entire body to the correct position, requiring the feet and legs subsystems. Then reach out with their arm/hand to intersect with the baseball traveling toward them and finally grasp and hold on to the ball when it reaches their hand. Without depth perception, it would be far more difficult to judge the baseball's path. With out proper body/arm/hand movement, which also requires visual feedback, it would be difficult to catch the ball. Finally, the hand grabbing the baseball at the proper moment and holding on to it requires tactile feel to accomplish the end result.

Obviously, there's a lot of behind the scenes coordinated work going on to make a seemingly simple event come off without a hitch. Consider the human brain at a high level:
The **Cerebrum** controls language, reasoning, motor skills, tactile senses, vision and sound.
The **Cerebellum** is responsible for balance, posture and coordination of movement. Note, the Cerebellum contains most of the neurons in the brain.
The **Limbic System** controls the emotions, motor and sensory functions, hunger, thirst, fear ,memory, converting short term memory into long term memory and spacial relationships.
The **Brain Stem** is responsible for breathing, heartbeat, movement, hearing and vision.

A little more on the distribution of neurons in each of the brain sections. The
Cerebral cortex has approximately 21 to 26 billion neurons. The Cerebellum is said to have about 100 billion.

To put a little more perspective on the brain's activity: Although it may seem an unsophisticated movement, finger tapping is a simple, rhythmic movement that utilizes a surprisingly large amount of brain activity in the cerebellum and frontal lobes.

Bob will need something close to this structure, sans the hunger, thirst, emotion, heartbeat, breathing. Bob may need some other items like automatic diagnostics.
The other interesting aspect of the human brain is it can grow more brain cells if needed. So, either Bob should have a huge number of spare unused brain cell equivalents available or a means of making more, if needed.

e) Recognition System

The non-motion functions of the brain are visual images, sounds or smells and combinations thereof. As previously discussed, vision, sound and smell can be an element used in coordination with movement. Aside from the movement aspect, during normal activity the incoming images, sounds or smells are recorded and compared to previously recorded ones. This would enable recognizing of familiar smells, sounds or images. Thus enabling recognizing faces, for example. This also occurs for recognizing sounds and smells and combinations of all three. This of course is being learned by the infant and continues forever as the child becomes an adult. This activity does provide the stimuli for the human brain activity during rest (I.e. Dreaming).

Bob would, just as any human record and compare images, smells and sounds. Bob would not be necessarily dreaming , however, he may review images, sounds or smells after the fact. Could this constitute a stimulus for image, etc., research or "thinking" ? Could Bob come up with a new image, etc. based on review of previous images in order to solve a problem, if asked to do so by his trainer or co-worker? Could this be considered artificial creativity? Could Bob part take in a "brain storming" session with his coworkers? They may, one day, be all androids.

Chapter 5 ... What is Learning ?

Learning can be defined in general, obviously by trying something and examining the results. Then evaluating whether the results of a given action were good or bad. However, what constitutes a good or a bad result also has to be learned. Finally, the information "learned" must be saved to be used later, so it won't have to be re-learned in the future.

At a high level, a good or bad result has to be based on feedback as a result of an action. Therefore when "learning" is modeled, it must include the means of producing an action and a means of measuring the results.

This chapter will present conceptual background information on numerous design disciplines which are envisioned to be needed in the implementation of this self learning system.

a.   Simple Closed Loop System

A closed loop system is a control system where the output of the system is monitored and fed back into the system as an input to the system. As an example of a closed loop system, consider walking in a straight line drawn on the ground. If your eyes are open, you get feedback on where you are stepping, and can correct any missteps as you walk. It is easy to walk straight correctly.

A more generic and universal way to look at a closed loop system is *something is proportional to something else multiplied by a constant of proportionality over time*. As some odd, but accurate examples: when a person is hungry, they eat until full, one drives a car until you arrive at the desired destination, one searches for a lost item, until it is found.

A common and simple example, could be a closed loop system. That is, a system with feedback to control the output could be considered a "learning" system. A closed loop system, a system with feedback, could also be described as a "goal-oriented" system, a system which seeks a goal. With either one of these views of the feedback system, the results the designer wants to achieve is predefined or designed into the system at the beginning.

Lets take an electric motor with a tachometer attached to the shaft of the electric motor. The "controller" would have as it's output a variable speed control for the electric motor. The controller would increase the electrical power output to the motor to increase the speed of the motor or conversely decrease the electrical power output to decrease the speed of the motor. The tachometer connected to the motor spins, as the motor spins, produces more voltage output the faster the motor spins and less when the electrical motor slows down. To control the speed of the motor a "set point" is introduced to tell the motor the desired speed. This set point is what determines for the feedback system what is good or bad.

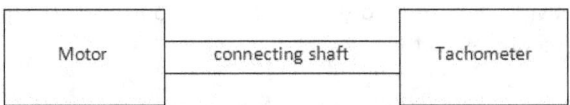

Figure 4 - Motor with tachometer feedback

The controller tries to match the set point at all times. If the motor speed goes faster than the set point allows, the power output of the motor is reduced. If the motor speed is too slow, the controller increase the output power to the motor and the motor speeds up. The amount of the up or down adjustment is usually small so as not to vary the motor's speed wildly. Obviously, if the load, that is the mechanical resistance the motor was being presented varied, the controller would have to provide varying power to the electric motor to maintain the same speed under the changing load conditions.

As previously stated, this closed loop system (electric motor, tachometer and controller) works to satisfy a goal, the "set point". In order to remove the operator set point from the system another level of control must be added. Another "closed loop" system on top of the first closed loop system needs to set an even higher "goal" if you will.

To build on the electric motor, tachometer and controller subsystem, lets say the motor is powering a pump, which is pumping a chemical into a mixer. The mixer requires a certain amount of this chemical per hour or per weight or a percentage of the total chemicals in the mixer or some other criteria. Therefore, there could be a higher level feedback loop system controlling the "set point" of the motor closed loop subsystem.

Let's say this higher level closed loop system is controlling several motor subsystems, each pumping a particular chemical into the mixer. And for this example the chemicals are controlled based on a percentage of the overall mix. The higher lever controller would set each of the "set points" for each of the pump motors based on the desired percentage mix. A chemical analyzer examines the resulting mixture and provides feedback to the higher level controller to adjust the percentages on the mix.

The higher level controller could have a setting to produce so many gallons of the chemical mix per hour or day or week, etc.. This could be yet a higher level closed loop system which adjusts based on the sales projections for the product.

To complicate this hypothetical system further, let's say at the final output of this mixing system, the effectiveness of the chemical product itself is evaluated and adjustments to the percentages of each chemical are modified. This would represent an even higher level of a feedback loop system.

Logically, all the parameters (i.e. settings, motor speed settings, chemical percentages) "learned" during the production of this chemical compound would be saved and used again if the same compound were needed again.

b.   Neural Network Systems

At a high level a neural network has some characteristics similar to a feedback or goal oriented system. A single neural net would have several "weighted" inputs which present their summation to the net and the net's would go to true if the summation of the input exceeded a threshold. The "weighted" importance of the individual inputs is obtained during "training" thru feedback based on samples presented to the input. An interesting feature of the neural net is that it inputs work in parallel, that is all the input information is processed at the same time.

Let's examine a simple neural network designed to recognize printed alphabetic characters. A system like this would be presented numerous examples of printed alphabetic characters in order to "learn" the attributes of each printed character. Therefore the system utilizes "feed back" of patterns and assigns (learns) the importance of the different characteristics.

Take 3 neural nets, each designed or taught to recognize the printed capital letters "A", "B", "C" respectively. For simplicity the neural net would have a 5x7 matrix view of the printed letters.

Letter A:

		X		
	X		X	
X				X
X	X	X	X	X
X				X
X				X
X				X

Figure 5 - Letter A - 5x7 matrix

Letter B:

X	X	X		
X			X	
X			X	
X	X	X		
X			X	
X			X	
X	X	X		

Figure 6 - Letter B - 5x7 matrix

Letter C:

	X	X		
X			X	
X				
X				
X				
X			X	
	X	X		

Figure 7 - Letter C - 5x7 matrix

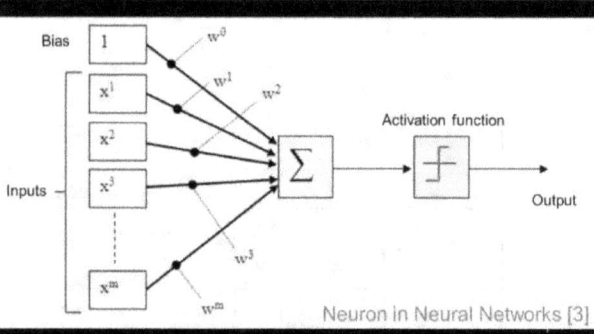

Neuron in Neural Networks [3]

Figure 8 - Simple Neural Network Diagram

Each neural net would have 35 inputs (5x7)and would be presented these patterns respectively in order to learn and recognize in the future it's letter. Taking a closer look at the individual neural net, each of it's 35 inputs is weighted. That is some inputs have a greater importance in detecting it's assigned alphabetic letter.

The relative importance of each of the inputs is determined using a re-iterative "learning" process. There are numerous methods utilized with Neural Nets to adjust the weights of the inputs. We'll use a simple method for this example where the initial value of the 35 inputs "weights" start off at some number (usually a random number). Variations of the desired letter are presented and each input is rated for it's importance in identifying the letter.

Letter A:

Sample 1:

		X		
	X		X	
X				X
X	X	X	X	X
X				X
X				X
X				X

Sample 2:

	X	X	X	
X				X
X				X
X	X	X	X	X
X				X
X				X
X				X

Sample 3:

		X		
	X		X	
	X		X	
X	X	X	X	X
X				X
X				X
X				X

Figure 9 - Samples Letter A

For each iteration the value (importance) of the individual pixel is adjusted. Lets say if the pixel is present that input value increases by 10%, conversely when no pixel is present the input value decreases by 10%. The more iterations, the more accurate the input becomes. Finally, the sum of the inputs is presented to the neural net and if the value exceeds a "threshold" it will "fire" and produce a positive output. The threshold would generally be anywhere from 50 to 70 percent certainty.

A closer look at the learning (teaching) process could be shown as follows:
For each of the thirty five inputs, for each iteration:

	Yes	Add 10% of multiplier	
Pixel			Multiplier (range 0 - 100%)
	No	Subtract 10% of multiplier	

Figure 10 - Neural Network Learning Diagram

This calculation occurs on all inputs at the same time, "parallel processing", every time a sample is presented in the learning or training mode.

As mentioned after the training is completed, the pixel being present or not (1 or 0) is multiplied by the "multiplier" for each of the 35 inputs as a sample character is presented. All thirty five inputs are added together and if the sum reaches a threshold, the desired character is considered recognized.

Neural net systems can get much more complicated than this simple example. For character recognition, the grid would probably be more like 500X700 pixels. Further, multiple layer neural net systems using hundreds of neural nets could be used for more complex problems, facial recognition for example.

While other Neural networks use different methods of calculating the weights of the individual inputs during training, self learning neural networks learn as they go. They could use for example, a search algorithm with heuristic to come up with the weighted inputs. More on this later.

c.   **Expert Systems**

From Wikipedia, the free encyclopedia:

> *An expert system is a computer system that emulates the decision-making ability of a human expert. Expert systems are designed to solve complex problems by reasoning through bodies of knowledge, represented mainly as **if-then-else** rules. An expert system is divided into two subsystems: the interface engine and the knowledge base. The knowledge base represents facts and rules. The inference engine applies the rules to the known facts to deduce new facts. The first expert systems were created in the 1970s and then proliferated in the 1980s. Expert systems were among the first truly successful forms of artificial intelligent (AI) software.*

An Expert Systems is a joint effort between a Subject Matter Expert (a SME) and a programmer that are focused on a particular problem. Diagnosing a car malfunction or trying to fix a ceiling light fixture or even diagnosing an illness are good examples of expert systems. A subject matter expert would work together with a software programmer and put together a series of questions in order to guide the user to a possible result or diagnosis or recommendation. It is like an on-line trouble shooting guide.

Let's try a simple example for diagnosing a ceiling fixture which has stopped working. The expert system may start by asking: has the ceiling light recently stopped working ? Yes/No. The user may respond Yes. The system might then ask: does the light work if a known good light bulb is inserted in the ceiling fixture ? Yes/No. The user may respond No. The system may then ask: please check the circuit breaker box for the light circuit. Does the light work now ? Yes/No. The user responds Yes. The system then says: Problem solved.

After each Yes/No answer the expert system would take a different path in it's logic and ask a different question of the user. Diagnosing someone's illness or a malfunction in a car would be much more involved and the system would have to ask a lot more question before reaching any conclusion or recommendation.

To further complicate expert systems, fuzzy logic is added. Fuzzy Logic imitates the way of decision making in humans that involves intermediate possibilities between digital values YES and NO. The inventor of fuzzy logic, Lotfi Zadeh, observed that unlike computers, the human decision making includes a range of possibilities between YES and NO, such as:

| CERTAINLY YES |
| POSSIBLY YES |
| CANNOT SAY |
| POSSIBLY NO |
| CERTAINLY NO |

Figure 11 - Fuzzy Logic

The fuzzy logic works on the levels of possibilities of input to achieve the definite output. An expert system using fuzzy logic, for example, could be an extremely simple temperature regulator that uses a fan might look like this:

IF temperature **IS** very cold **THEN** stop the fan

IF temperature **IS** cold **THEN** turn down the fan

IF temperature **IS** normal **THEN** maintain level

IF temperature **IS** hot **THEN** speed up the fan

IF temperature **IS** very hot **THEN** turn the fan on high

The fuzzy logic introduces intermediate states between just Yes and No. The **if-then-else** construct is still needed to select between very cold and cold, neutral, hot and very hot. Fuzzy logic software is used mostly in simple applications.

From - Guru99.com

See the below-given diagram. It shows that in fuzzy systems, the values are denoted by a 0 to 1 number. In this example, 1.0 means absolute truth and 0.0 means absolute falseness.

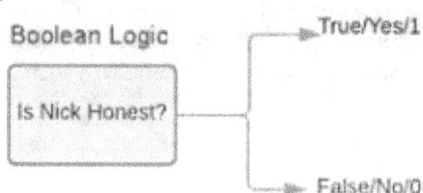

Figure 12 - Boolean Logic Diagram

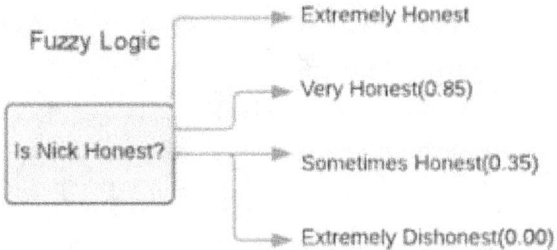

Figure 13 - Fuzzy Logic Diagram

Product	Company	Fuzzy Logic
Anti-lock brakes	Nissan	Use fuzzy logic to controls brakes in hazardous cases depend on car speed, acceleration, wheel speed, and acceleration
Auto transmission	NOK/Nissan	Fuzzy logic is used to control the fuel injection and ignition based on throttle setting, cooling water temperature, RPM
Auto engine	Honda, Nissan	Use to select gear based on engine load, driving style, and road conditions.
Copy machine	Canon	Using for adjusting drum voltage based on picture density, humidity, and temperature.
Cruise control	Nissan, Isuzu, Mitsubishi	Use it to adjusts throttle setting to set car speed and acceleration
Dishwasher	Matsushita	Use for adjusting the cleaning cycle, rinse and wash strategies based depend upon the number of dishes and the amount of food served on the dishes.

Figure 14 - Companies using Fuzzy Logic

Elevator control	Fujitec, Mitsubishi Electric, Toshiba	Use it to reduce waiting for time-based on passenger traffic
Golf diagnostic system	Maruman Golf	Selects golf club based on golfer's swing and physique.
Fitness management	Omron	Fuzzy rules implied by them to check the fitness of their employees.
Kiln control	Nippon Steel	Mixes cement
Microwave oven	Mitsubishi Chemical	Sets lunes power and cooking strategy
Palmtop computer	Hitachi, Sharp, Sanyo, Toshiba	Recognizes handwritten Kanji characters

Figure 15 - More Companies using Fuzzy Logic

Now consider a fully trained neural network which results in a set of weighted inputs. These weighted inputs could be the same as though they were set up by a question and answer session with an expert system. Perhaps single, fully trained neural networks could be viewed as mini expert systems of limited scope. Expert system software with fuzzy logic may be useful in programming some limited scope dedicated neural network functions.

d.   Search Techniques

There are two types of searching techniques :

Uniformed search methods include a Binary Search, Linear Search, Depth-First Search and the Breadth-First Search. These algorithms do not know anything about what they are searching. Uninformed searching is a brute force approach to solving a problem and usually take a lot of time to reach an end point.

Informed search would be where the search algorithm has some method of selecting a path to test rather than randomly choosing. A **Heuristic** search is an informed search technique. A heuristic value tells the algorithm which path will provide the solution as early as possible. The heuristic function is used to generate this heuristic value. Different heuristic functions can be designed depending on the searching problem.

i.   Binary Search Algorithm

The binary search is a limited but efficient algorithm even though it is a blind, brute force method. The binary search algorithm does however require the data being searched to be organized. So a binary search would efficiently search data which are numeric or alphabetic by dividing it in half, then half again etc..

A simple example of a binary search would be to look up the meaning of a word. So, we would need to have a database organized alphabetically:

Entry	Word	Meaning
1	Adam	Person's name
2	Aardvark	Animal
3	Apple	Red Fruit
4	Balloon	To get bigger
5	Basket	Container
↓		↓
997	Watermelon	Big fruit
998	Xylophone	Musical instrument
999	Yellow	Color of a banana
1000	Zebra	Four legged animal

Figure 16 - Binary Search Database Example

In this example the binary search algorithm would be given a word by the user and would present the user with the meaning it found in the database, if any.

Let's say they requested the meaning of the word **watermelon.** Instead of searching each word and comparing for a match to the word watermelon, the algorithm would divide the entire database in half. Let's say there were 1000 entries in our database, the algorithm would go to entry number 500 and compare the first letter of that word if it is alphabetically before or after the first letter of the desired word, in this case the letter "w".

Since it was after the letter in the 500th entry, the algorithm would divide the lower half of the database in half, picking the 750th entry. And again the "w" is after the 750th entry, the remaining database is divided again in half and so on. Each time dividing the remaining database in half until it finds the word starting with "w". The algorithm would then check the rest of the word for a match.

Once the requested word is found the meaning of the word, in this case "Big fruit" is returned to the user. The algorithm task is complete and it is awaiting the next word request.

ii. Linear Search Algorithm

An **uninformed** linear search can be depth first or breadth first type of search.

The **depth first** algorithm would search the first path encountered all the way to a conclusion before checking the adjacent path. The breadth first would check all adjacent paths before moving down a level and checking all the next level's paths.

The following diagram represents the decision flow of a **depth first** search algorithm.

Each clear circle represents a decision point which has not been tried. A colored circle means it has been tried or evaluated by the algorithm. The up arrow lines shows back tracking path.

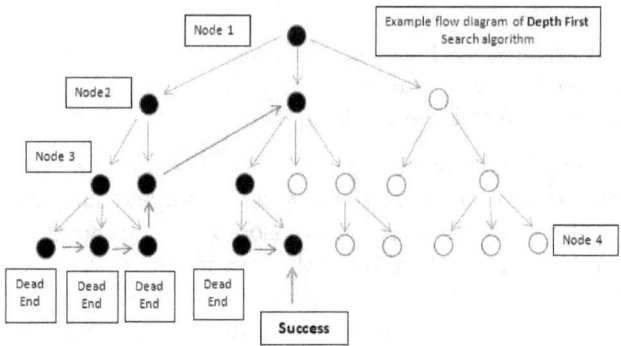

Figure 17 - Depth First Search Example

In the **depth first** example, the algorithm started at **Node 1**, picking the left most path, bringing it to **Node 2**. At **Node 2**, it selected the left most path, bringing to **Node 3**. From there it went to the left most path down to **Node 4**, where it encountered a **Dead End**. The up arrow indicates the algorithm is back tracking to select the second choice in **Node 4**. This turned out to be a **Dead End**, causing the algorithm to select the third choice on the **Node 4** level, also a **Dead End**. The algorithm then backed up to the **Node 2** level and selected the second choice path. It continued to **Node 3** and then **Node 4** where it again encounter a **Dead End**. Finally, it tried the second choice in the **Node 2** level and found a solution to the problem, **Node 3** then **Node 4,** right most path, **Success**. In this example of the **depth first** algorithm visited 11 decision points before finding the solution

The **breadth first** algorithm will visit each decision point in the **Node 2** level, then each decision point at the **Node 3** level. The algorithm will continue on the **Node 4** level and so on, until every decision point has been checked. The following diagram represents the decision flow of a **breadth first** search algorithm.

Each clear circle represents a decision point which has not been tried. A colored circle means it has been tried or evaluated by the algorithm.

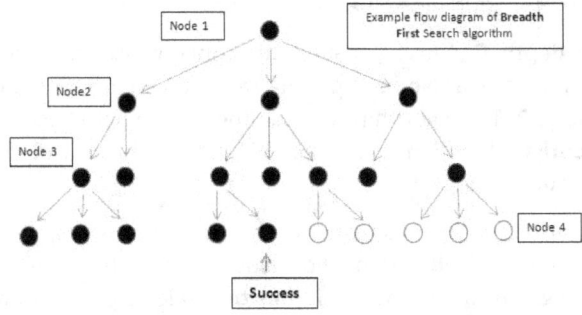

Figure 18 - Breadth First Search Diagram

In the **breadth first** example, the algorithm started at **Node 1**, picking the left most path, bringing it to **Node 2**. Then selecting path 2 from **Node 1,** checking the second decision point in **Node 2**, then on to the third path to **Node 2**. The algorithm checks all the paths in **Node 3**. Finding nothing, goes on to check all paths in **Node 4**. At the fifth path on **Node 4**, the algorithm finds the solution, **Success**. In this example of the **breadth first** algorithm visited 16 decision points before finding the solution.

The following diagram represents the decision flow of a **depth first** search algorithm using a **heuristic function**.

Each clear circle represents a decision point which has not been tried. A colored circle means it has been tried or evaluated by the algorithm.

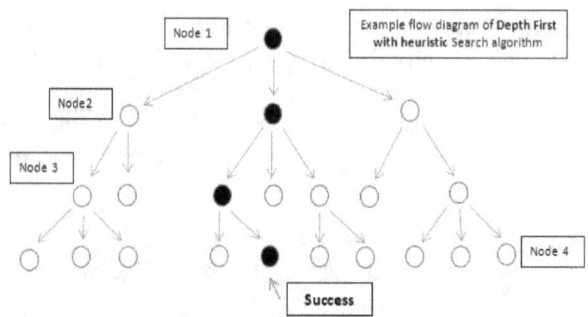

Figure 19 - Heuristics Search Diagram

In the **depth first** using a **heuristic function** example, the algorithm starts at **Node 1**, picking the middle path, bringing it to **Node 2**. The algorithm selected the middle path because the **heuristic function** evaluated this path as most likely to lead to success.

The algorithm then picks path 1 from **Node 2** to **Node 3**, since path 1 was judged by the **heuristic function** to be most promising. Finally, the **heuristic function** selects path 2 from **Node 3** to **Node 4**, which results in the solution, **Success**. In this example of the **breadth first with heuristic function** the algorithm visited 4 decision points before finding the solution.

Using a **heuristic function** does not guarantee the algorithm won't hit a **dead end** during the search. In that case, the depth first algorithm would back up (as was shown in the depth first example) and pick the next untried path at that Node and use the heuristic function as before.

A simple example is the sliding puzzle:

The **Sliding puzzle** is a popular puzzle that consists of N tiles where N can be 8, 15, 24 and so on. In our example N = 8. The puzzle is divided into sqrt(N+1) rows and sqrt(N+1) columns. Eg. 15-Puzzle will have 4 rows and 4 columns and an 8-Puzzle will have 3 rows and 3 columns. The puzzle consists of N tiles and one empty space where the tiles can be moved. Start and Goal states of the puzzle are shown below. The puzzle can be solved by moving the tiles one by one in the single empty space and thus achieving the Goal configuration.

Initial State: Goal State:

1	2	3
	4	6
7	5	8

1	2	3
4	5	6
7	8	

Figure 20 - Start position and end goal position of tiles

The tiles in the initial (start) state can be moved into the empty space in a particular order and thus achieve the goal state.

Rules for solving the puzzle:

> The empty space is basically swapped an adjacent tile. The empty space can only move in four directions:

> - Up
> - Down
> - Right
> - Left

Let's follow a simple depth first algorithm for solving this problem.

For simplicity lists are shown as 1,2,3,4,5,6,7,8,**B** where the tile number is listed starting from row 1, column 1 to column 3, row 2, column 1 to 3 and row 3 column 1 to 3. The letter B is the blank position.

The **available list** would have every conceivable combination of the 9 values in it, which means the available list has 9 factorial minus 1 or 362879 entries. The algorithm can only take an entry from the available list if it is logical to use from the current position. In other words, only an entry which legally moves one tile from the current position can be considered as a possible next move. Therefore, there can only be two, three or four possible moves from any given state to the next.

The first entry in the **visited list** is the starting state.

Step 1:

	Visited		Legally Available
1	1,2,3,B,4,6,7,5,8	1	1,2,3,4,B,6,7,5,8
		2	1,2,3,4,B,6,7,5,8
		3	1,2,3,7,4,6,B,5,8

Available 1

1	2	3
4		6
7	5	8

Available 2

	2	3
1	4	6
7	5	8

Available 3

1	2	3
7	4	6
	5	8

Figure 21 - Step one of search

The algorithm picks the first value from the available list and moves it to the visited list.

	Visited
1	1,2,3,B,4,6,7,5,8
2	1,2,3,4,B,6,7,5,8

Figure 22 - Results of step 1 of search

Step 2:

The next step would now have the following visited and legal available lists. The available entries are legally available moves from previously visited, entry 2. The first entry in the available column is not usable, since it is already in the visited list. So the algorithm would choose the second available entry.

	Visited		Legally Available
1	1,2,3,B,4,6,7,5,8		
2	1,2,3,4,B,6,7,5,8	1	1,2,3,4,B,6,7,5,8
		2	1,2,3,4,6,B,7,5,8
		3	1,2,3,4,5,6,7,B,8

Available 1*

1	2	3
4		6
7	5	8

Available 2

1	2	3
4	6	
7	5	8

Available 3

1	2	3
4	5	6
7		8

Figure 23 - Step 2 of search

* Already visited, therefore not available even though it is a legal position

Again, the first entry in the available column is not usable, since it is already in the visited list. So the algorithm would choose the second available entry and put it on the visited list.

	Visited
1	1,2,3,B,4,6,7,5,8
2	1,2,3,4,B,6,7,5,8
3	1,2,3,4,6,B,7,5,8

Step 3:

	Visited		Legally Available
1	1,2,3,B,4,6,7,5,8		
2	1,2,3,4,B,6,7,5,8		
3	1,2,3,4,6,B,7,5,8	1	1,2,B,4,6,3,7,5,8
		2	1,2,3,4,6,B,7,5,8
		3	1,2,3,4,6,8,7,5,B

Available 1

1	2	
4	6	3
7	5	8

Available 2*

1	2	3
4	6	
7	5	8

Available 3

1	2	3
4	6	8
7	5	

Figure 24 - Step 3 of search

* The second entry in the available list appears in the visited list, therefore it is not available. The algorithm would then choose available entry #1 and move it into the visited list.

	Visited
1	1,2,3,B,4,6,7,5,8
2	1,2,3,4,B,6,7,5,8
3	1,2,3,4,6,B,7,5,8

Figure 25 - Results of step 3 of search

Step 4 , Step 5, etc. will follow the same procedure and on and on till it reaches the goal state or runs out of available moves.

If it runs out of available moves in the current path, it will back up one step and try the next available untried entry. Of course the visited entry would be removed when it backs up, putting that entry back in the available list.

If following that new path results in the same dead end, the algorithm will back up another level and try the next untried available move at that step. It will continue in this manner until it finds the goal state or has no more options. If there is a solution to the problem, this search methodology will find it eventually.

Literally, by the **process of elimination**, this algorithm will find a solution, if one exists. This is not, however, an efficient algorithm. It is a blind, brute force method of solving a puzzle.

The sample puzzle presented was a very simple problem. Real life problems can be much more complex and therefore the efficiency of the algorithm would become important. If a method of evaluating a given choice being better than the others could be used at each step, it can vastly improve the algorithm's efficiency.

i.   Heuristic Algorithm

The word heuristic in every day situations is perceived as meaning a **rule of thumb** or **common sense**.

From Wikipedia, the free encyclopedia:
> *A **heuristic technique**, often called simply a **heuristic**, is any approach to problem solving, learning, or discovery that employs a practical method, not guaranteed to be optimal, perfect, logical, or rational, but instead sufficient for reaching an **immediate goal**. It is a function that ranks alternatives in search algorithms at each branching step based on available information to decide which branch to follow. Heuristics underlie the whole field of Artificial Intelligence and the computer simulation of thinking, as they may be used in situations where there are no known algorithms.*

The heuristic is used instead of trying every possible state in solving a problem, since heuristic approach should find the result faster. The key to using a heuristic approach to solve a

46

problem, lies in having a methodology for evaluating the current state of the problem with respect to the end state of the problem.

From Wikipedia, the free encyclopedia:

> *In computer science, **A*** (pronounced "A star") is a computer algorithm that is widely used in pathfinding and graph traversal, which is the process of finding a path between multiple points, called "nodes". It enjoys widespread use due to its performance and accuracy. However, in practical travel-routing systems, it is generally outperformed by algorithms which can pre-process the graph to attain better performance, although other work has found A* to be superior to other approaches.*
>
> *Peter Hart, Nils Nilsson and Bertram Raphael of Stanford Research Institute (now SRI International) first published the algorithm in 1968. It can be seen as an extension of Edsger Dijkstra's 1959 algorithm. A* achieves better performance by using heuristics to guide its search.*

An important feature of the A* algorithm is that it keeps a track of each visited node which helps in ignoring the nodes that are already visited, saving a huge amount of time. It also has a list that holds all the nodes that are left to be explored and it chooses the most optimal node from this list, thus saving time not exploring unnecessary or less optimal nodes.

The open list contains all the nodes that are being generated and are not on the closed list. Each node explored after it's neighboring nodes are discovered is put in the closed list. The neighbors are put on the open list this is how the nodes expand. Each node has a pointer to its parent so that at any given point it can retrace the path to the parent. Initially, the open list holds the start (Initial) node. The next node chosen from the open list is one with the lowest cost as calculated by the number of nodes already traversed plus the distance from the end point.

The following is based on an internet article by Ajinkya Sonawane - Sept. 15, 2018 and also a school project I had during an AI class.

A* uses a combination of heuristic value (**h-score**: how far the goal node is) as well as the **g-score** (i.e. the number of nodes traversed from the start node to the current node).

In the 8-Puzzle problem, the **h-score** is the number of misplaced tiles by comparing the current state and the goal state. The **g-score** will remain as the number of nodes traversed from the start node to get to the current node. The **f-score** equals the   **g-score** plus the **h-score**.

The initial state of the puzzle has **h-score** = 4 since there are four tiles out of place. The **4, 6, 5** and **8**. The **g-score** = 0 as the number of nodes traversed from the start node to the current node is 0.

Initial State:

1	2	3
	4	6
7	5	8

Goal State:

1	2	3
4	5	6
7	8	

Figure 26 - Start and End State

Now evaluate the three possibilities in step G=1. The software will test moving the #4 tile, the #1 tile and the #7 tile.

In the first case, moving the #4 tile results in 2 tiles out of place. Therefore, H=2 and F=3. Moving the #1 tile in the second case results in 4 tiles out of place, H=4 and F=5. The third case moves tile #7 resulting in 4 tiles out of place, H=4 and F=5. Therefore the winner of round one is case number one, moving the #4 tile.

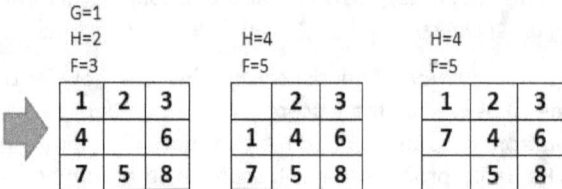

Figure 27 - Results after step 1 of heuristic search

Next, evaluate the three possibilities in step G=2. The software will test moving the #2 tile, the #6 tile and the #5 tile.

In the first case, moving the #2 tile results in 4 tiles out of place. Therefore, H=4 and F=6. Moving the #6 tile in the second case results in 3 tiles out of place, H=3 and F=5. The third case moves tile #5 resulting in 1 tiles out of place, H=1 and F=3. Therefore the winner of round two is case number 3, moving the #5 tile.

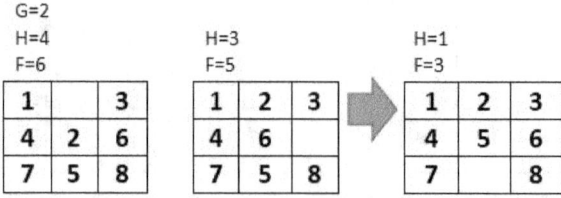

Figure 28 - Results after step 2 of heuristic search

Next, evaluate the two possibilities in step G=3. The software will test moving the #7 tile and the #8 tile.

In the first case, moving the #7 tile results in 2 tiles out of place. Therefore, H=2 and F=5. Moving the #8 tile in the second case results in no tiles out of place, H=0 and F=3. Therefore the winner of round three is case number 2, moving the #8 tile.

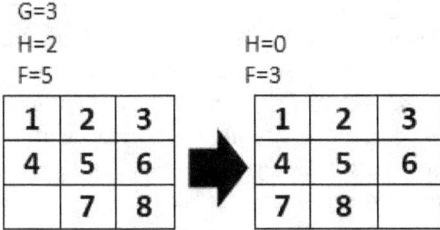

Figure 29 - Results after step 3 of heuristic search

When no tiles are out of place, H=0. The goal state has been reached and the solution is complete.

It would seem likely in an extremely large and complex neural network system, a sizeable number of neural networks would have a need to be dynamically trainable. Examples are an android climbing up a mountainside or standing still in a wind storm or not falling over on the deck of a ship at sea.   An A-Star heuristic algorithm should be useful in neural net dynamic learning. Learning in real time, as it is happening.

e.   Artificial Neural Networks

The typical neural network used in more complex problems today is referred to as artificial neural networks. They typically use multiple layer's of neural nets and different algorithms to **learn**. There are supervised and unsupervised, static and dynamic types of neural networks. A huge amount of information can be found online on this subject. A lot of it is not easy to understand and can be a bit arcane. The following excerpts from Wikipedia gives a reasonable overview.

From Wikipedia, the free encyclopedia

> ***Artificial neural networks (ANN)* or *connectionist systems*** *are computing systems vaguely inspired by the biological neural networks that constitute animal brains. The neural network itself is not an algorithm, but rather a framework for many different machine learning*

algorithms to work together and process complex data inputs. Such systems "learn" to perform tasks by considering examples, generally without being programmed with any task-specific rules. For example, in image recognition, they might learn to identify images that contain cats by analyzing example images that have been manually labeled as "cat" or "no cat" and using the results to identify cats In other images. They do this without any prior knowledge about cats, for example, that they have fur, tails, whiskers and cat-like faces. Instead, they automatically generate identifying characteristics from the learning material that they process.

An ANN is based on a collection of connected units or nodes called artificial neurons, which loosely model the neurons in a biological brain. Each connection, like the synapses in a biological brain, can transmit a signal from one artificial neuron to another. An artificial neuron that receives a signal can process it and then signal additional artificial neurons connected to it.

In common ANN implementations, the signal at a connection between artificial neurons is a real number, and the output of each artificial neuron is computed by some non-linear function of the sum of its inputs. The connections between artificial neurons are called 'edges'. Artificial neurons and edges typically have a weight that adjusts as learning proceeds. The weight increases or decreases the strength of the signal at a connection. Artificial neurons may have a threshold such that the signal is only sent if the aggregate signal crosses that threshold. Typically, artificial neurons are aggregated into layers. Different layers may perform different kinds of transformations on their inputs. Signals travel from the first layer (the input layer), to the last layer (the output layer), possibly after traversing the layers multiple times.

The original goal of the ANN approach was to solve problems in the same way that a human brain would. However, over time, attention moved to performing specific tasks, leading to deviations from biology.

Artificial neural networks have been used on a variety of tasks, including computer vision, speech recognition, machine translation, social network filtering, playing board and video games and medical diagnosis.

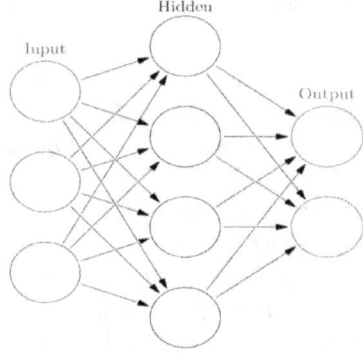

Figure 30 - Neural Network with hidden layer

An artificial neural network is an interconnected group of nodes, inspired by a simplification of neurons in a brain. Here, each circular node represents an artificial neuron and an arrow represents a connection from the output of one artificial neuron to the input of another.

The Hidden Layer is the part of the artificial neural network that does the learning. It takes information from the Input Layer and learns to match them up with the desired output.

From Wikipedia, the free encyclopedia

Types of artificial neural networks

Artificial neural networks have many variations. The simplest, static types have one or more static components, including number of units, number of layers, unit weights and topology. Dynamic types allow one or more of these to change during the learning process. The latter are much more complicated, but can shorten learning periods and produce better results. Some types allow/require learning to be "supervised" by the operator, while others operate independently. Some types operate purely in hardware, while others are purely software and run on general purpose computers.

Applications

Because of their ability to reproduce and model nonlinear processes, Artificial neural networks have found many applications in a wide range of disciplines.

Application areas include system identification and control (vehicle control, trajectory prediction, process control, natural resource management), quantum chemistry,general game playing, pattern recognition (radar systems, face identification, signal classification, 3D reconstruction, object recognition and more), sequence recognition (gesture, speech, handwritten and printed text recognition), medical diagnosis, finance (e.g. automated trading systems), data mining, visualization, machine translation, social network filtering and e-mail spam filtering.

Artificial neural networks have been used to diagnose cancers, including lung cancer, prostate cancer, colorectal cancer and to distinguish highly invasive cancer cell lines from less invasive lines using only cell shape information.

Artificial neural networks have been used to accelerate reliability analysis of infrastructures subject to natural disasters and to predict foundation settlements

Artificial neural networks have also been used for building black-box models in geoscience: hydrology, ocean modelling and coastal engineering, and geomorphology.

Artificial neural networks have been employed with some success also in cyber security, with the objective to discriminate between legitimate activities and malicious ones. For example, machine learning has been used for classifying android malware, for identifying domains belonging to threat actors and for detecting URLs posing a security risk. Research is being carried out also on ANN systems designed for penetration testing, for detecting bot-nets, credit cards frauds, network intrusions and, more in general, potentially infected machines.

Now for the arcane. The type of **algorithms used in the artificial neural networks** can vary widely. The following are excerpts taken from Wikipedia, the free encyclopedia

a. Group method of data handling

The Group Method of Data Handling (GMDH) features fully automatic structural and parametric model optimization. The node activation functions are Kolmogorov-Gabor polynomials that permit additions and multiplications. It used a deep feedforward multilayer perceptron with eight layers. It is a supervised learning network that grows layer by layer, where each layer is trained by regression analysis. Useless items are detected using a validation set, and pruned through regularization. The size and depth of the resulting network depends on the task.

b. Convolutional neural networks

A convolutional neural network (CNN) is a class of deep, feed-forward networks, composed of one or more convolutional layers with fully connected layers (matching those in typical Artificial neural networks) on top.

Examples of applications in computer vision include DeepDream and robot navigation.

c. Long short-term memory

Long short-term memory (LSTM) networks are RNNs that avoid the vanishing gradient problem. That is, LSTM can

learn "very deep learning" tasks that require memories of events that happened thousands or even millions of discrete time steps ago.

In 2003, LSTM started to become competitive with traditional speech recognizers.In 2007, the combination with CTC achieved first good results on speech data. In 2009, a CTC-trained LSTM was the first RNN to win pattern recognition contests, when it won several competitions in connected handwriting recognition. In 2014, Baidu used CTC-trained RNNs to break the Switchboard Hub5'00 speech recognition benchmark, without traditional speech processing methods. LSTM also improved large-vocabulary speech recognition, text-to-speech synthesis,for Google Android, and photo-real talking heads. In 2015, Google's speech recognition experienced a 49% improvement through CTC-trained LSTM.

LSTM became popular in Natural Language Processing. LSTM can learn to recognise context-sensitive languages. LSTM improved machine translation, LSTM combined with CNNs improved automatic image captioning.

d. Deep reservoir computing

Deep Reservoir Computing and Deep Echo State Networks (deepESNs) provide a framework for efficiently trained models for hierarchical processing of temporal data, while enabling the investigation of the inherent role of RNN layered composition.

e. Deep belief networks

A deep belief network (DBN) is a probabilistic, <u>generative model</u> made up of multiple layers of hidden units. It can be considered a <u>composition</u> of simple learning modules that make up each layer.

A DBN can be used to generatively pre-train a DNN by using the learned DBN weights as the initial DNN weights.

Backpropagation or other discriminative algorithms can then tune these weights.

f. Large memory storage and retrieval neural networks

Large memory storage and retrieval neural networks (LAMSTAR) are fast deep learning neural networks of many layers that can use many filters simultaneously. These filters may be nonlinear, stochastic, logic, non-stationary, or even non-analytical. They are biologically motivated and learn continuously.

LAMSTAR has been applied to many domains, including medical and financial predictions, adaptive filtering of noisy speech in unknown noise, still-image recognition, video image recognition, software security and adaptive control of non-linear systems. LAMSTAR had a much faster learning speed and somewhat lower error rate than a CNN based on ReLU-function filters and max pooling, in 20 comparative studies.

These applications demonstrate delving into aspects of the data that are hidden from shallow learning networks and the human senses, such as in the cases of predicting onset of sleep apnea events, of an electrocardiogram of a fetus as recorded from skin-surface electrodes placed on the mother's abdomen early in pregnancy, of financial prediction or in blind filtering of noisy speech.

LAMSTAR was proposed in 1996 (A U.S. Patent 5,920,852 A) and was further developed Graupe and Kordylewski from 1997 to 2002. A modified version, known as LAMSTAR 2, was developed by Schneider and Graupe in 2008.

g. Stacked (de-noising) auto-encoders

The auto encoder idea is motivated by the concept of a good representation. For example, for a classifier, a good representation can be defined as one that yields a better-performing classifier.

In order to make a deep architecture, auto encoders are stacked. Once the stacked auto encoder is trained, its output can be used as the input to a supervised learning algorithm such as support vector machine classifier or a multi-class logistic regression.]

h. Deep stacking networks

A deep stacking network (DSN) (deep convex network) is based on a hierarchy of blocks of simplified neural network modules. It was introduced in 2011 by Deng and Dong. The structure of the hierarchy of this kind of architecture makes parallel learning straightforward, as a batch-mode optimization problem. In purely discriminative tasks, DSNs perform better than conventional DBNs.

i. Tensor deep stacking networks

This architecture is a DSN extension. It offers two important improvements: it uses higher-order information from covariance statistics, and it transforms the non-convex problem of a lower-layer to a convex sub-problem of an upper-layer. The basic architecture is suitable for diverse tasks such as classification and regression.

j. Spike-and-slab RBMs

The need for deep learning with real-valued inputs, as in Gaussian restricted Boltzmann machines, led to the spike and slab RBM (ssRBM), which models continuous-valued inputs with strictly binary latent variables. Similar to basic RBMs and its variants, a spike-and-slab RBM is a bipartite graph, while like GRBMs, the visible units (input) are real-valued. The difference is in the hidden layer, where each hidden unit has a binary spike variable and a real-valued slab variable. A spike is a discrete probability *mass at zero, while a slab is a <u>density</u> over continuous domain; their mixture forms a <u>prior</u>.*

An extension of spike and slab <u>RBM</u> called micro spike and slab <u>RBM</u> provides extra modeling capacity using additional terms in the <u>energy function</u>. One of these terms enables the model to form a <u>conditional distribution</u> of the spike variables by <u>marginalizing out</u> the slab variables given an observation.

k.   Compound hierarchical-deep models

*Compound hierarchical-deep models compose deep networks with non-parametric <u>Bayesian models</u>. <u>Features</u> can be learned using deep architectures such as DBNs,DBMs, deep auto encoders, convolutional variants, ssRBMs, deep coding networks, DBNs with sparse feature learning, RNNs, conditional DBNs, de-noising auto encoders. This provides a better representation, allowing faster learning and more accurate classification with high-dimensional data. However, these architectures are poor at learning novel classes with few examples, because all network units are involved in representing the input (a **distributed representation**) and must be adjusted together (high <u>degree of freedom</u>).*

To explore more of ANN and see a large cross section of the variations used in the learning algorithms refer to Wikipedia **Artificial Neural Networks**.
The development of the artificial neural network field is still taking place and will probably continue for some time. Suffice to say **we have the technology** to make neural networks work.

f.   Object Oriented System Architecture

Object Oriented architecture uses a programming language based on modeling a collection of entities known as objects. The objects model the characteristics of real world or conceptual entities. Each object defined would have internal procedures designed to mimic the behavior of the object it represents. The object could modify it's own internal data or external data. An Instruction and/or data is sent to the object in the form a message.

As a simple example let's say a TV remote control was an object. Separate commands could be sent to it, one to change the channel and one to change the volume setting. The commands may look like **change_channel (26)** and **change_volume (75)**. These would be the internal procedures (commands): change_channel and change_volume.

Another example, could be a conceptual entity such as a **window** on a computer screen. The window could be sent a message to display the contents of a spreadsheet. While another window on the same computer user's screen could have a game playing in it. Another could have a word processor working on a document in it. Each of the user's windows or **objects** operate independent of each other. Each with their own data set and behavioral characteristics. Multiple instantiations of a type of object are called a **class**. So in this example the **window object** is a **class**.

Excepts from *www.tutorialspoint.com*

Object
- An **object** *is a real-world element in an object oriented environment that may have a physical or a conceptual existence. Each object has*
- *Identity that distinguishes it from other objects in the system.*
- *State that determines characteristic properties of an object as well as values of properties that the object holds.*
- *Behavior that represents externally visible activities performed by an object in terms of changes in its state.*

Objects can be modeled according to the needs of the application. An object may have a physical existence, like a customer, a car, etc.; or an intangible conceptual existence, like a project, a process, etc.

Class

*A **class** represents a collection of objects having same characteristic properties that exhibit common behavior. It gives the blueprint or the description of the objects that can be created from it. Creation of an object as a member of a class is called instantiation. Thus, an object is an **instance** of a class.*

The constituents of a class are:

- *A set of attributes for the objects that are to be instantiated from the class. Generally, different objects of a class have some difference in the values of the attributes. Attributes are often referred as class data.*
- *A set of operations that portray the behavior of the objects of the class. Operations are also referred as functions or methods.*

Example

Let us consider a simple class, Circle, that represents the geometrical figure circle in a two dimensional space. The attributes of this class can be identified as follows:

- *X coord, to denote x coordinate of the center*
- *Y coord, to denote y coordinate of the center*
- *a, to denote the radius of the circle*

Some of its operations can be defined as follows:

- *findArea(), a method to calculate area*
- *findCircumference(), a method to calculate circumference*
- *scale(), a method to increase or decrease the radius*

Object Oriented languages:

Smalltalk is an object-oriented was created as the language in underpinning the **new world** of computing exemplified by **human computer symbiosis**. It was designed and created by the Learning Research Group (LRG) of Xerox PARC by Alan Kay, Dan Ingalls, Adele Goldberg, Ted Kaehler, Scott Wallace, and others during the 1970s. **Smalltalk** was one of many object-oriented programming languages based on the language **Simula**. Virtually all of the object-oriented languages that came after: **Flavors, CLOS, Objective-C, Java, Python, Ruby**, and many many others were influenced by **Smalltalk**.

g. Goal Oriented Systems Architecture

Goal Oriented Systems Software Architecture can provide the overarching structure for the design of an autonomous android since it can provide the top level goals to control the myriad of underlying functional systems.

Self-management (computer science) from Wikipedia, the free encyclopedia

> **Self-Management** *is the process by which computer systems shall manage their own operation without human intervention. Self-Management technologies are expected to pervade the next generation of network management systems.*
>
> *The growing complexity of modern networked computer systems is currently the biggest limiting factor in their expansion. The increasing heterogeneity of big corporate computer systems, the inclusion of mobile computing devices, and the combination of different networking technologies like WLAN, cellular phone networks, and mobile ad hoc networks make the conventional, manual management very difficult, time-consuming, and error-prone. More recently self-management has been suggested as a solution to increasing complexity in cloud computing.*

Currently, the most important industrial initiative towards realizing self-management is the Autonomic Computing Initiative (ACI) started by IBM in 2001. The ACI defines the following four functional areas:

- **Self-Configuration**: Automatic configuration of components;
- **Self-Healing**: Automatic discovery, and correction of faults; automatically applying all necessary actions to bring system back to normal operation[3]
- **Self-Optimization**: Automatic monitoring and control of resources to ensure the optimal functioning with respect to the defined requirements;
- **Self-Protection**: Proactive identification and protection from arbitrary attack.

While the following is paper cited discusses the application of Goal-Oriented Software approach to business, it is certainly more than applicable to Artificial Intelligence Systems.

Excerpts from Goal-Oriented Approach to Self-Managing Systems Design
Steven J. Bleistein and Pradeep Ray. School of Information Systems, Technology, and Management, University of New South Wales, Kensington 2052, NSW, Australia

Self-managing systems are computing systems that can manage themselves according to high-level objectives that system administrators define for them.
IBM refers to these kinds of systems as **"autonomous computing"** citing four aspects self-management: **self-configuration, self-optimization, self-healing, and self-protection**. For the present, these aspects are treated as distinct solutions of self-managing systems. In the future, IBM foresees gradual steps of evolution's in self-managing systems in which the distinctions become blurred, ultimately becoming general properties of "self-maintenance" in a general architecture.

IBM envisions the application of this technology in highly strategic ways, beyond the vision of network *maintenance.   The IBM Autonomic Computing Manifesto cites examples in mass distribution retail, health care, and e-services. Self-managing Systems are in fact interactive collections of autonomous "elements", or more precisely, autonomous agents. These agents manage their internal behavior and relationships with other autonomous agents in accordance with policies established by human managers or other agents. The behaviors of autonomous elements are programmed in higher-level, goal-oriented terms.   This leaves agents with the responsibility to resolve details of behaviors, relationships, and connections on the fly. Therefore, design of self-managing system in part depends on agent-oriented design methodologies.*

Functional vs. Non-Functional Requirements

Most software engineers are familiar with the concept of functional requirements (FR). In modeling system requirements, non-functional requirements (NFR) are often just as important, and can make or break a project depending on how well they are satisfied. **NFR**s *are defined as system requirements that are not tied to any particular functionality. Examples of non-functional requirements include system-wide requirements such as performance, accessibility, and security. In this respect, modeling* **NFR**s *is relevant to modeling requirements of self-managing systems in that the four aspects cited by IBM of* **self-configuring, self-optimizing, self-healing,** *and* **self-protecting** *are all in fact non-functional requirements by this definition.*

Hard-goals vs. Soft-goals

Goals are distinguished between hard and soft.   The distinction between the two is that of measurability. Hard-goals are satisfied when measurable criteria have been achieved.   Soft-goals however, have no concrete measurable criteria. Their status is thus "fuzzy." Instead of discussing soft-goals in terms of being satisfied, which implies an absolute completeness regarding goal achievement, soft-goals are "satisfied." Similar to the use of the term in domain of decision support systems, the nuance is that a goal has been satisfied to a degree that is "good enough" to accept.

*The importance in the distinction between hard and soft goals is that **NFR**s are often represented in terms of soft-goals. Moreover, high-level business goals are very "soft" in nature. Therefore, representation and treatment of soft-goals is important in design of self-managing systems.*

An example of a Goal Oriented System

***A Goal-Oriented Approach for the Generation and Evaluation of Alternative Architectures** - Gemma Grau, Xavier Franch - Universitat Politècnica de Catalunya (UPC) c/ Jordi Girona 1-3, Barcelona E-08034, Spain.*

The Home Service Robot Case Study

This case study is based on the problem statement of the prototype of a Home Service Robot (HSR) for daily services. A Goal-Oriented Approach for the Generation and Evaluation of Alternative Architectures

The HSR is a prototype that supports the following daily home services:

- *Call and Come (CC).*
- *User Following (UF).*
- *Security Monitoring (SM).*
- *Tele-presence (TP).*

The robot sends the remote user a map of the house and the user can command the robot to move to a specific position. In addition, the robot can send captured images to the remote PDA for surveillance. In order to provide those services, the HSR has the following hardware components: a Single Board Computer that controls the peripherals; a Front Camera to allow face recognition, user tracking, security monitoring and tele-presence; a Ceiling Camera to do map building and self-positioning; 8 SL Microphones to interpret speaker commands and locate its specific position; a Structured Light Sensor to detect obstacles and recognize footsteps; an Actuator to allow the HSR movement; an LCD display to show information; a Wireless Lan to communicate to the Home Server; and, finally, a Speaker to generate sound. Goal dependencies stating functional requirements, e.g. the User depends on the HSR for the goals Come when called and Accidents are avoided. – Softgoal dependencies stating high-level non-functional requirements.

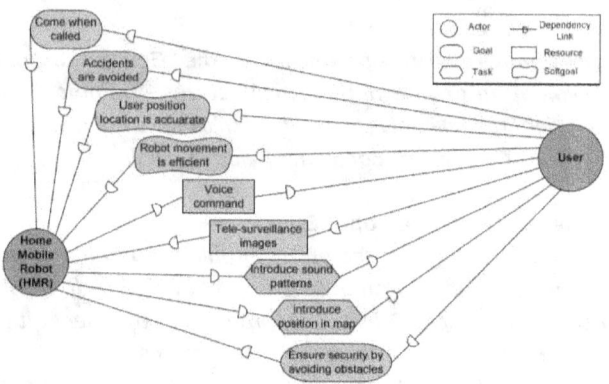

Fig. 1. Excerpt of an *i** model for the HSR case study

Figure 31 - Goal System Diagram

Chapter 6 ... Building Blocks

A *Conceptual Model is a way of representing a particular concept, or set of concepts, that helps people understand or simulate the subject of that model.*

The descriptions and discussions presented are not intended to dictate the implementation of various functions, but to describe their behavior.

Some general statements regarding **Neural Nets**:

➤ With judicious selection of inputs, the correct algorithms for calculating the relative importance of these inputs and sufficient training, almost any set point or goal objective can be reached. Note the output of the Neural Net generally represents a high probability of being correct but is not necessarily 100% accurate. The accuracy generally depends on the amount of training the Net has gone thru I.e. the number of examples used during training

➤ Neural Nets can be divided into layers where numerous lower level Nets provide input to the next level of Nets thus allowing more complex goals to be achieved.

➤ The Neural Network functionality may be easier to understand when it is used to recognize images or control movement (I.e. pattern recognition). This functionality is exactly same, however, when used to analyze more abstract situations like deciding between two or more choices so long as the inputs are properly chosen.

➤ Regarding inputs to Neural Networks, basically too many inputs are actually OK, since as training occurs, the superfluous inputs will be unused and go to zero in their significance. This is very much like the brain where the less used links between neurons simply fade away when not used.

➢ The tricky aspect of selecting the Neural Net inputs is to be sure to include those inputs which are important in finding solutions. Obviously, the more significant the inputs are to reaching the solution the more accurate or higher confidence the solution will have. It is sort of like solving a series of robberies in the neighbourhood. Look for the common elements between the different crimes. The crimes were all done at night, all the witnesses said the crook wore a green hat, they walked with a limp, etc..

a.   Self Learning Cell - the smallest unit

Starting at the bottom is always a good place to start. In all likely hood there may be many millions of these cells within the system. Just as the molecule is the smallest physical unit of an element or compound, so too is the proposed self learning cell. The **self learning cell** is the smallest unit in this system and the building block for virtually the entire system. This cell contains an artificial neural network, multiple sensory inputs, a memory, outputs and a goal or set point input.

The **sensory inputs** are the inputs to the neural network. The **output** is a signal level or a message indicating the inputs have satisfied the goal or set point. The **memory** or internal private data storage is to remember the settings from various learning sessions. The **goal** input, of course stimulates the output to try to satisfy the new goal or set point. A goal rather than a set point is used, since the set point implies the output matches the set point exactly or not, where a goal is like fuzzy logic and only needs to be close enough.

The goal can be a message telling the memory which data set to use (or find a new data set) and causes the output to change. The goal input message may also contain the search criteria for the weighting of the inputs and mode. The basic **modes** of operation could be training, dynamic learning or use previously learned input weighting.

Remembering how any feedback system works, the set point tells the system to change it's output in order to allow the inputs to satisfy the criteria and balance the system.

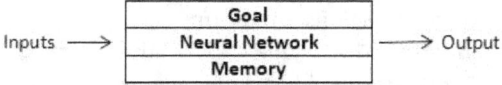

Figure 32 - Self Learning Cell Diagram

The input goal command could contain the following parameters:

    i.   **memory set** - which memory set to use or new set.

    ii.   **mode of operation** - learning, dynamic learning, or use previously learned weighting

    iii.   **search criteria** - specifies the type of learning algorithm to use*

    iv.   **set point** - the desired signal level

* there are many possible learning algorithms. **A** thru **K** is cited in the **algorithms used in the artificial neural networks** in **Chapter 5.e**

Note also, that the input processing would be parallel, therefore virtually the total response time from the output stimuli to the changes of the input would be due to   the external element being manipulated.

b.   **Self Learning Objects,**

This system uses a construct called an **object.** The idea of an object comes from object oriented software architecture. For our purposes, the object is a stand alone, independent collection of hardware and software. The object is the next level up of a control element envisioned in this system is akin to a software object. Like software objects, these objects can be of a certain class, have a unique identity, contain functions, produce outputs, have inputs, maintain internal databases, have access to external databases, have access to external resources (like internet access, visual and sound input or outputs), etc.. The objects communicate with other objects in the system by sending and receiving messages.

Unlike software objects, however, these objects should be implemented mostly in hardware for speed.

i.   **Type of object** - class or category of object based on the functions it performs

ii. **Identity** - The unique name of the object

iii. **Functions** would include any specific algorithms, but probably consist mostly of a number of **self learning cells**.

iv. **Outputs** - may be messages or actual signal to some type of actuator(s)

v.   **Inputs** - may be messages or signals from sensors

vi. **Internal database** - to keep track of a task in progress or remember previously performed tasks

vii. **Access to external databases** - could be considered part of the object's output or simply record keeping for multiple objects to share or a cloud database

viii. **Access to external resources** - video cameras, sensors, sound, or internet access in order to complete it's tasks

After the self learning cell, the object is the most common element in this system, numbering in the order 100's of thousand. The objects or functions in this system are compartmentalized, meaning there is an object for just about everything from moving muscle equivalents to individual subjects like running or playing chess or learning history or retrieving memories (visual, audio, smells), etc..

A look at the messages that the object may receive would give further insight into it's generic functionality. The message would contain an operational **mode**, which could be:

i.   **Learn** - the training mode for the object and the self learning cells

ii. **Run** - The normal operational mode after the training has been done

iii. **Configure** - Define the external sources are needed for this object:
    a. **Internet**
    b. **Balance**
    c. **Audio**
    d. **Visual**
    e. **GPS**
iv. **Algorithm** - Use a special algorithm for Neural Network calculations, **A** through **K**
v. **Maintenance** - As the name implies, this mode allows for:
    a. **Initialization** - Clear the memory, clear the self learning cell (Neural Networks). Like a complete reboot
    b. **Rename** - To install the object's reference name so it will be addressed by higher level objects in command messages
    c. **Retrieve Memory** - Upload the object's database
    d. **Diagnostics** - Perform internal testing of the objects major components to assure proper operation

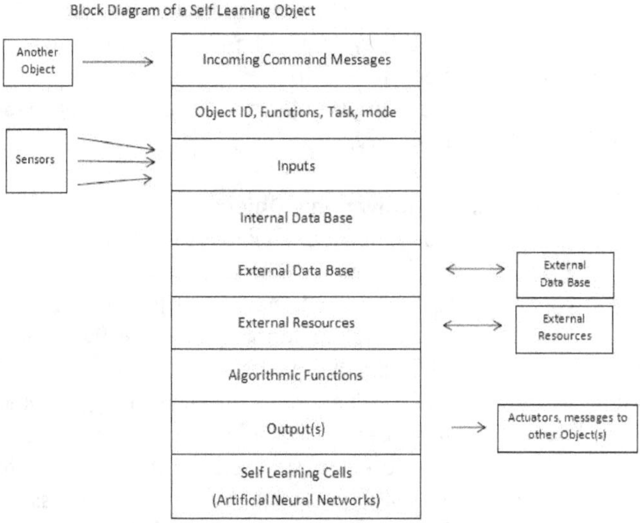

Block Diagram of a Self Learning Object

Figure 33 - Block Diagram of Self Learning Object

Types of Objects in the self learning system are as follows:

a. **Motion Object**
Motion Objects interface with the actuators or muscle-equivalents to move the framework of the humanoid. These are responsible for all physical motion.

b. **Visual Object**
Visual objects, as the name implies, handles the vision input for the system.

c. **Audio Object**
Audio objects takes care of, listening and speaking

d. **Memory Object**
Memory Object manages the data going into and out of the long term memory of the system.

e. **Object Manager**
The object manager is responsible for determining if more objects are required to be put on line. In order for this system to handle learning in various subjects, there needs to be a large inventory of unused or unassigned objects. These can be made active if necessary.

f. **Motivational Object**
Motivations are what gets things moving. It is the force that initiates, guides, and maintains goal-oriented behaviors. It is what causes the system to take action. In this system **Motivational Objects** get things going.
The Motivational Object is a type, which provides messages to many other objects, which themselves are closer to the actual task being performed. The Motivational Object, while not necessarily initiating the specific task, is responsible for doling out the commands to get the task done.

g. Goal Object

The goal objects are the top cops in the system. They keep the system in-line. There are a number of goal objects providing multiple layer goals management. Some Goal Objects would be responsible for maintenance, like internal self testing of functions. Another would be responsible directing one or more motivational object to get an overall goal accomplished.

h. Analysis Object

The analysis object is used in critical thinking simulation.

Chapter 7 ... Motion Objects

Starting at the bottom of the control system is the motion objects and the control of the framework of the android. These objects handle the movement of the android and therefore one motion object is needed for each *muscle* in the frame. The system must be *aware* of it's physical structure in order to know the hierarchical control layout. That is a *map* of the structure is part of the control scheme.

> *Just like the bones song:*
> the toe bone connected to the foot bone,
> and the foot bone connected to the ankle bone,
> and the ankle bone connected to the leg bone.

a.   Legs

There would be an object for each toe in the foot, one for the ankle, one for the calf, one for the thigh, hip, torso, stomach, back, etc., then each joint in each finger and the hand, then the wrist, forearm, upper arm, shoulder and on and on. There would be motion objects for the neck movement, a number of objects for the facial muscles equivalents, etc.. The human adult has around 200 bones and around 600 muscles and so would the human android.

Let's take one of the five toes on one foot. An input message to the toe could be, bend the toe down 10%, and messages to the adjacent four toes, bend 8% through 2% respectively. These five messages would be coming from the foot object.
Inputs to these toes would be coming from torso's balance and pressure sensors from each toe and sensors to feedback the position of the toe. The Neural Nets would be calculating and adjusting the amount of torque to apply to maintain the desired bend.

The ankle motion object would be receiving a message from the calf object and translating it to the foot object, which sends the five messages to the individual toes. Perhaps the ankle is being told to have a 10% bend up with a 5 degree offset from center. The ankle has 6 directions of motion, up down, twist left or right and yaw and pitch, left or right.

All of these motion objects would be trying to achieve the desired position and maintain them with feedback from the individual angular displacement sensors, individual pressure sensors and balance sensors. As you would image the messages to the individual motion objects are flowing down passing through each of the higher level motion objects involved in the overall movement.

In each motion object, the command that passes through, adds or modifies or expands the messages to accomplish the task. All of this activity starts at a higher level with a command let's say to walk along a path or climb up the stairs. The decision to walk or run or climb the stairs comes from an even higher object and would be instigated by some type of motivation object. The motivation object would send a command to the walk motion object or the run motion object or the climb stairs motion object, etc..

The high level motion objects would obviously have inputs, including visual reference so as to not bump into anything or **see** where the next stair is, or stay on the path, etc.. The top level walking object would also have GPS information as one of it's inputs.

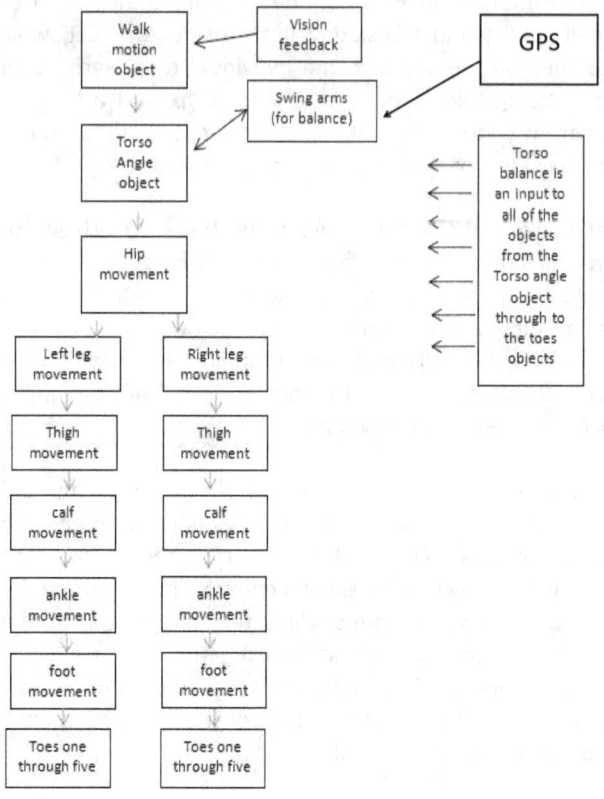

Figure 34 - Communications between objects when walking

This is where learning via artificial neural networks really come into play. Generally, each of the **movement objects** shown above has feed back inputs from the overall balance of the android, their internal individual angular displacement sensor(s), skin surface pressure and internal individual mechanical load pressure sensors.

In each of the motion objects, these inputs would provide performance feedback to their internal individual neural networks. Allowing the weighting calculations to take place and eventually provide the correct weighting values for each of those inputs.

In each of the **movement objects'** internal databases, the required weighting factors are saved and played back during the entire leg movement cycle as the walking occurs. The leg movement cycle for walking would of course repeat until the android has walked to it's destination.

In each of the movement objects, from the Torso down to the toes, there would be a level of dynamic (learning as you go) adjustment. The individual motion object's neural networks weighting calculations would account for irregularities in the terrain or wind conditions, etc. as the android walks. In other words, after the initial training of all the motion objects involved in **walking** during ideal conditions, there would still be a need for dynamic adjustments to occur on the fly, in order to account for less than ideal or *real world* conditions, which always occur.

Logically, aside from the walk object, there would be a run object, a climb up the stairs object, climb down the stairs object, to name a few. Every activity which is significantly different from the others, would have it's own top level object. Significantly different in this case means a movement which is more than dynamic adjustments could compensate for.

b.   Arms

Modeling the hand and arm of the android is not unlike the leg. Let's look at the hand. To model the human hand literally, one would have to consider three joints in each of four fingers and two joints in the thumb as **individual finger objects**. A total of 14 finger sections objects per hand. They of course are connected to the hand itself and then on to the wrist.

> The finger bone segments (the scientific name for the finger bones is **carpals**) are named **phalanges**:
> • Tip segment: **Distal phalanx.**
> • Middle segment: **Middle phalanx.**
> • Bottom segment: **Proximal phalanx**

The motion objects shown below represent the control for each of the individual bones within the hand and wrist. As was described in the discussion of the leg, the control messages go from the wrist object to the hand object, then to the four fingers joint 3, thumb joint 2. From there each finger joint sends command messages to the joint 2 and joint 1 for the thumb. Finally to the finger joint 1 in each of the four fingers.

As in the leg, the wrist, hand and finger motion objects from a higher level would be sent a command to grab a glass of water for example. The wrist would rotate appropriately, the hand would send the command to the individual finger to grasp the glass correctly, applying the correct amount of pressure so as to not drop the glass, but not break either. The individual finger sections would rely on their pressure sensors to provide the feedback. These commands would most likely also have feedback from the vision object at a higher level to verify the motion is being performed correctly.

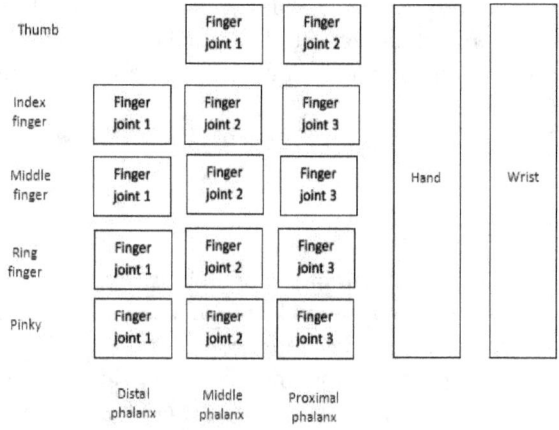

Figure 35 - Hand and finger motion objects

The forearm, upper arm and shoulder objects would be next as we move up the chain. The command message flows, as expected is from a higher level, to the shoulder, upper arm, forearm and on to the wrist.

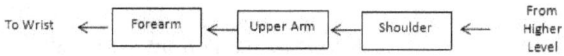

Figure 36 - Arm to wrist motion objects

c.   Other

The neck, jaw, head plus some 42 muscles in the face would each have their own motion object controlling those movements.

Chapter 8 ... Visual Objects

The visual processing system is controlled by the visual object. The visual processing system consists of two cameras, each with some focus mechanism to allow for "zooming-in" on distant objects and capable of distance measurement via triangulation of these cameras.

Major functions of the visual system and visual object would include **object recognition**, **object avoidance**, **distance measurement**, **video image storage** and **video image comparisons**.

Taking closer look at the distance measurement task of the visual object, let's say we want to catch a baseball that was hit high into right field. A higher level object would tell the visual object to find and track the baseball. Using the visual object's two cameras, it would focus on the high flying baseball coming this way. It calculates the baseball's flight path, trajectory, speed and anticipated arrival time and position.

The cameras continually adjust their angle (the android's head may have to turn to maintain visual contact) and focus to track the baseball throughout it's flight. As the cameras track the baseball, the visual object is sending position data to other subsystems which after some translation, will be sent to numerous motion objects. That information is moving the entire android into the anticipated position to catch the baseball.

If climbing stairs, the visual object is focused on the next step for the android to step on. If typing on a keyboard the visual object is focusing on the desired keyboard button the android is typing. In both these examples, the visual object is passing, positioning information the rest of the objects, which is needed it at that time.

The visual system has background tasks running at all times, **obstacle avoidance, object recognition** and **automatic image storage**.

a.   **Automatic image storage**

Automatic image storage would occur whenever the scene (field of view) changed to a significant extent. For example, when a change of 5 percent of the pixels in a scene occurs, the image would automatically be saved. These images could be culled by other maintenance objects in order to free up storage space periodically.

b.   **Object recognition**

Recognizing objects in an image (the field of view) is a background process, constantly running and maintaining a dynamic table of previously identified physical objects. The physical objects are compared to known physical objects in the image database. The known images were previously learned during training sessions. Like flash cards in grammar school, the android would learn the word apple with a picture of an apple next to it, a fire truck with a picture, a cow, etc..

c.   **Obstacle avoidance**

Obstacle avoidance is another background process, which is constantly looking to identify any object on a collision course with the android. Whether the object is moving towards the android or the android is approaching the obstacle, it should be avoided. Examples of this would be walking into a wall or a piece of flying debris heading at the Android. If an obstacle is detected a message would be sent to a higher object in order to determine a course of action. When an obstacle is detected and the collision course is calculated, the decision of what to do about it actually rests with the analysis object. This will be discussed later in the chapter on analysis objects.

Further, the visual object would allow other objects in the system to use the video processor as a resource.

d.   Video image comparisons

Image comparisons would be a part of a higher level task. That is, a command may be sent to the visual object to look for a missing cat. The cat's image would be sent or linked with the command. The visual object would notify the requesting higher level object when the desired item has been located.

e.   Distance measurement

Distance measurement would be another command sent by a higher level object in the system. This command would also include an image or a link to the image of the item whose distance is being measured. Obviously, once the item has been located, it distance measurement would also be passed to the requesting higher level object.

f.   Common sensory functions

Each of the sensors (**eyes, ears, nose, balance** and **skin**) are available to all other objects in the system as a resource and can be used directly as inputs if required. A common function of all resources in the system is that they each run background processes to monitor for dangerous or problem situations. The potential problems the sensors may encounter include:

➢   The **video processor** observing fire or smoke
➢   The **odor processor** detecting fire or smoke
➢   The **audio processor** detecting a loud and or unidentified noise(s)

- ➢ The **balance processor** detecting an unusual or unexpected shift in the android's balance.
- ➢ The **tactile processor** detecting excessive pressure or unexpected temperatures anywhere on the android's skin

When any of these situations are detected the object or objects where the resource is contained will immediately send an alarm type message to a higher level object to determine the appropriate course of action(s). This message is considered a high priority system level **interrupt**, since it could mean an emergency situation which must be immediately dealt with.

Chapter 9 … Audio Objects

The audio object and audio processing system would provide the ability to "**hear**" and "**produce**" sounds. The sound input information would be in stereo, allowing limited directional information to be interpolated. Additionally, the audio processing system would have a background program listening for loud or unusual sounds, like dishes breaking or a siren or any unexpected sound.

a.   Automatic sound storage

When a sound is detected, the background processing compares it to previously identified sounds, like a cow's moo, bird's chirping or dog's barking, etc.. An active database is maintained of all currently active and known sounds in the area and the source of each of the sound's relative locations around the android. The sounds detected can be stored simultaneously with corresponding image information.

b.   Loud noise detection

Detected sounds are measured for frequency, amplitude and would additionally be evaluated for the direction of the source of the sound. A higher level **analysis object** may determine the android's head should be turned to better determine the direction of the sound or to literally **see** what the source of the sound might be. Detecting an unexpected sound would cause an **interrupt** to be sent up the control chain where it can be evaluated, since it may represent a danger or some kind.

The audio object is controlling the foreground program which is listening for words. When a word is detected it is compared to previously stored words. Each word received is evaluated to determine the meaning. The words would be passed on to an **analysis object** as a message, to decide any action which should be taken based on the word or string of words received. Certain key words would be built in like **Stop** , the **Android's name** (which could be changed), **Come here**, **Sit down**, **Sleep**, **Wake up**, etc..

The audio object would of course handle speaking the words which are sent to it from higher level analysis object responsible for outgoing communications. The audio object could also handle producing sounds as well, like a bird chirping, a dog barking, etc.. These **speaking** or **produce a specific known sound** command messages would include the words or sound to be sent out, the volume and perhaps in the case of words some inflections, surprise, emphasis, question, etc..

Chapter 10 ... Odor, Balance and Tactile Sensor Objects

The odor detector, the balance sensor and the tactile sensor (the **skin**) have corresponding objects, the **odor object**, the **balance object** and the **tactile object**. In each case, they run background programs, which are constantly detecting odors, the android's balance and sensing skin pressure and ambient temperatures.

a.   Odor detection

The simulated **nose** is responsible for detecting smells/odors. It records new smells and their intensity as they are detected. If commanded from a higher level the odor object can try to identify the source of the particular odor by viewing the source. After a database of smells are recorded, the background odor processor will send a message to a higher level if it detects something, which may indicate some type of problem or danger, for example, like smelling smoke or something burning or a strong unidentified odor.

b. Tactile sensor

The **tactile object** receives input from each of the motion objects (that's around 600 objects), since these objects contain the tactile sensors themselves. There certainly are a lot more tactile sensors in the skin. The following is from the internet:

> One square centimeter of human dermis has up to fifty receptor endings for pain, cold, warm and touch. These nerve endings can perceive one or more of senses like touch, pressure, cold, warm and pain. Some areas like tip of the fingers, lips and tongue have more touch receptor endings in human body. There are

many types of nerve receptor endings in human skin having specific functions of senses.

The tactile sensors or **skin** detects the amount of pressure in has on it, for example, picking up a glass full of water would produce, by necessity a certain amount of pressure on the finger tips in order that the glass does not slip but not so much pressure that the glass breaks.

Walking would produce more pressure on the bottom of the feet than elsewhere on the skin. Carrying a heavy package would produce more pressure in certain areas like the arms. Therefore the tactile object would only produce an **alarm state** if there were more pressure than expected any where on the android's body, while taking into account the current activities. The **interrupt message** would include the amount of pressure and the motion object or objects reporting this condition.

The **tactile sensor** also detects temperature. So, as in other sensory detection, if a higher than expected measurement is found, an alarm message would be sent to a higher level object. This **alarm message** would include the temperature value and the motion object or other objects reporting higher than usual temperature.

c.   **Balance sensor**

The **balance sensor** included in the **balance object** is constantly monitoring the android's balance, as in standing upright or at some degree from upright such that the android itself is not tipping over. The android could be in a controlled but not upright position on purpose, for example, bending over to pick something up or leaning forward into a strong wind and the like. An unexpected rapid change in balance may indicate a problem situation and would be forwarded up the chain to a higher level object for evaluation.

Chapter 11 … Memory Objects

All the data being sent in and out of the **permanent memory** would be processed by the **memory object**. Note, the permanent memory is the **common memory** in the system, not the local database within each object. The actual system's common database may partial reside in the android itself or in the cloud.

Obviously saving information in the resident database would allow for faster store and retrieve and available even when the cloud is not. Therefore, the memory object must make some determination of where certain pieces of information should be stored. It is possible as the android's memory fills, some information may be considered less vital and transferred to the cloud, thus freeing up local memory for higher priority images.

Recorded information such as instructions, commands given or words read would all be saved in text form in the permanent memory. A smell or sound would be literally be saved (for example, a **wav** or **avi** file) along with a description or any other reference at the time of recording. Preferable an image would be corresponding with the sound or odor and should all be saved and linked together.

Whenever an image is saved, it should have a corresponding **table of identified objects** (see chapter 8 visual objects). This table of identified items within the image should be saved along with the image.

Generally, **visual images** should be labeled with a unique name and be saved with or linked to the corresponding **sound**(s), **odor**(s), perhaps **temperature**, **date** recorded, **reason for recording**, etc.. The reason for recording may be a background routine recording or done in a command or part of a task or emergency, etc.. All the reason for recording the image information is to make the image's history as complete as possible, retrievable and traceable.

The image data is saved with it's table of identified objects, reference to sounds, odors, temperature, date and reason for recording so that it may be retrieved by any of these references as search criteria via the **memory object**. Further, searches of the data could be done by key words within text records like sentences for example.

Note, each of the identified items within the image data should have a link to detailed information about that item. Aside from a standard definition of the word, the item's classification (animal, mineral, plant, etc.), plus other information akin to an encyclopedia's information would be available.

The **memory processor** would have several background tasks continuously running, evaluating whether to save the information locally in the android or in the cloud. Another automatic task would be culling unimportant or duplicate image information, for example, if two images are nearly identical, one of them may be deleted freeing up space.

It may be required at some point down the road to eliminate some memory images due to lack of access. That is, if the image hasn't been accessed by the system in a long time, it may not be needed anymore and perhaps should be archived and then deleted.

Another background process running would be system fault detection. Should an error occur during any program execution in any of the objects in the system, a snapshot of software status at the moment the problem occurred would be saved in memory. Should an error occur, the technicians would be notified. Therefore, the **memory processor** should be accessible via a secure local link for software error snapshots and/or up loads by technicians.

Chapter 12 ... Object Manager

The object manager does just that, it managers the objects in the system. It's busiest area of concern is the motivational object, the number of which can constantly be expanded. The object manager maintains a list of active (online) objects of all types and is responsible for assigning new objects from the system's inventory of inactive (offline) objects.

The **self monitoring goal object** would send a command to the **object manager** to conduct routine tests, that is, it would initiate self tests for the active objects when they are idle. If an object fails it's self test, the object manager reports back to the goal object. If instructed by the **goal object**, the **object manager** would configure and put a replacement object online to replace the defective one. The **self monitoring goal object** would notify the Technician of the findings. The object manager further would maintain an inventory of defective objects.

Since all unassigned **objects** are the same in this system, the **object manager** can configure whatever is needed by specifying the type of **object**. This process is akin to an **FPLA** (field programmable logic array).

Chapter 13 ... Motivational Objects

Over time there would be probably in the order of hundreds of thousands, if not a million of motivational objects in the system. Motivational objects are the repository for various tasks in the system. It's envisioned the android would start off with perhaps a few hundred pre-programmed tasks. Just enough to allow new tasks to be recognized and created.

The need to please the trainer would require a motivational object to determine whether the trainer is expressing approval or disapproval of the actions the android has taken. Further, a rudimentary **walking** and **sitting** motivational object should be provided. Also, an understanding of a handful of voice command, like **Stop, Sit down**, **Sleep, Wake up** and the **Android's name**.

To further illustrate the motivational objects function, let's look at it from a very high level. Let's say there is a motivational object for **playing baseball**. It would contain a list of sub-motivational objects like to **run the bases, hit the baseball, catch the baseball** and so on. The **run the bases** motivational object would contain a list of other sub-motivational objects like **walking, running** and so on.

The **walking** and or **running** motivational object would contain a list of motion objects involved in moving the legs. Continuing along, as was discussed in the motion object chapter, the **hip** motion object all the way to the individual **toe** motion object would have a role in the walking.

Aside from the motivational object involved in walking or running, the different levels of motivational objects would use various resources in executing these tasks. As an example the **walking** and **running** motivation objects would need the **visual, GPS, sound, tactile** and **balance** resources.

The **hitting the baseball** motivational object would contain a list of other objects which control how to **stand when at bat**, how to **swing the bat**, keep your **eye on the ball**, and so on. Each of these objects would in turn contain lists of objects required to perform the individual tasks.

The **catch the ball** motivational object would have a corresponding list of required objects and sub-objects to carry out the task. Both hitting and catching objects would require resources along the chain of command in order to carry out the tasks at hand.

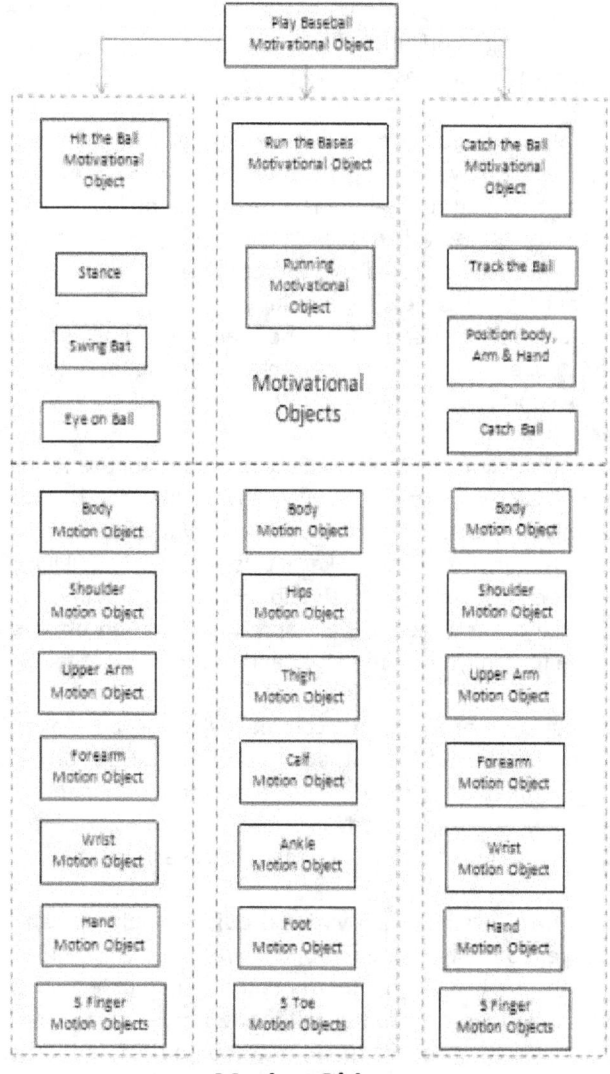

Motion Objects

Figure 37 - Playing baseball objects

New motivational objects would be created as the training of the android continues. As is envisioned in a self learning system, the knowledge base would increase as the android learns more tasks. Each task the android can perform has a corresponding motivational object to initiate the required procedures for that task.

Think of a child going to school learning their ABC's, it would work the same way with the self learning android. After first grade, the child probably has been taught many, many thousands of individual tasks. After completing grammar school the child would have been taught hundreds of thousands of tasks (reading, writing and arithmetic). Let's not leave out several years of all the standard school subjects: history, spelling and how to interact with fellow students, playing sports and games.

The typical motivational object would contain the required information to perform a given task. This information is probably a series of procedures that must be taken to complete the task. Let's take a closer look at walking, for example. The **walking motivational object** would send a sequence of messages to the appropriate **motion objects** to put one foot in front of the other and complete one step.

The sequence of messages (commands) could be move the left leg until the left foot has advanced a distance of 27 inches. Upon completion of the left leg's movement, the right leg is the sent a message to advance also 27 inches. If the goal location has not been reached, the sequence is repeated.

Actually, the **walking motivational object** would be sending messages to the **hip motion object**, which in turn would send messages to the **thigh motion object**. That object in turn would send messages to the **calf motion object**, which would in turn send messages to the **foot motion object**. From there, independent messages would be sent to each of the five **toe motion objects**. During these procedures, feedback would be provided from the visual, balance and GPS resources to all of the motivational and motion objects involved in the task.

The higher level motivational object would tell the **walking motivational object** to continue until the android reaches it's target location, possibly based on the GPS signal or a distance measurement from the visual processor.

Other motivational objects would include, facial expressions, (happy, sad), perhaps recite the Gettysburg Address or play a game of checkers. The motivation object sequence of commands for a given task are developed from the **learning object**, more on this later.

Chapter 14 … Goal Objects

Goals are like fuzzy commands. Doing well in school is a **goal**. Earning more money is another **goal**. Some goals may represent limits like the speed limit on a given street is a **goal** for all cars to strive for, but it also may mean don't exceed this limit. In this system, goal objects are there to define things to strive for but in some cases represent a limit.

The top level goal of this system is to **not harm any living thing**. Therefore, every physical movement is verified by this top level goal, so it is not exceeded. A second overall goal is not to displease the trainer/teacher. This is especially useful when the Android is less experienced, in it's "younger years". Another goal in this system is to constantly look to learn something new.

Some predefined goals in the system are:
- Top level (*the prime directive*) **do not harm any living thing**
- **Self testing** of and replacement of defective objects
- **Obstacle avoidance**

Automatic goals are generated by creating goal objects. These are to carry out specific goals. The goal object is very similar to a motivational object, except it covers a higher level. For example, the object goal could be **go to a meeting at 2pm in the conference room**. Obviously, this goal object would have several tasks in it, like collect the information or documents needed for the 2pm meeting, also at 1:45pm start walking to the conference room and take a seat at the conference table. In this example, the android would stop other non priority tasks at about 1:45pm in order to make it to the meeting on time.

A goal object is generally created by the audio object (together with an analysis object) when instructed by the trainer or someone else when requested to do something, other than what is covered in the motivational objects. The motivational objects are usually small and limited to **walk**, **stand up** or **sit down**. Unlike the motivational objects in this system, which are mostly **permanent**, the goal objects are usually **temporary**. When the goal has been reached, that goal object is deleted. For example, a "go to meeting" goal would be deleted once the meeting is finished. Note: Some goals would not be automatically deleted, such as a **play chess goal**.

Chapter 15 ... Learning Objects

The learning object has different methods of operation. In one of the methods, the **learning object** works in conjunction with the **analysis object** when a new physical item(s) or smell(s) or sound(s) is discovered. In another method, the **learning object** works together with the **audio, visual** and **tactile objects** when learning in a classroom environment, which would include discussions and reading. Learning as a result of physical movement, causes the learning object to literally program the sequence(s) required to complete a task within the motivational objects.

In each case, the learning object would compare what is being learned to previously learned material. The learning object would determine whether to add on to an existing category or start a new category.

A learning object might create a **play chess goal object** based on instructions from a trainer, a classroom environment or reading a book. There would be a **motivational object** for each move of each chess piece. The strategy to use during the game would come from the **critical thinking analysis objects**. Over time, the android's chess playing ability would improve. The android may or may not get to be a chess master depending on how much time the android puts into playing the game, not unlike most things in life.

Chapter 16 ... Analysis Objects

One way the analysis object makes a decision in the system is by analyzing the current image as viewed by the visual processor.

a.   Curiosity begets learning

As described earlier, in the **visual object's** description, the video processor in a background mode is automatically saving an image, if the image changes by about five percent. When this occurs, the visual processor identifies all items in the image. This information along with sound, smell and any tactile information are saved in the common system memory. That image and accompanying information, which has just been saved is also sent to the analysis object for review.

The **analysis object** takes this current image and accompanying information and compares it previously saved image and accompanying information, looking for as close as possible to a match. *Caution, **artificial neural nets** at work*. If a close match is not found, this image is considered new information, if a very close match is found no further processing with this image occurs.

Upon finding the **new information**, the analysis object may send (if someone else is present) a message to the **audio object** to ask for information about that which it does not recognize in the current view (image). The **audio object** would say to the nearby human for example, **what am I seeing ?** or if only one item in the scene (current view) is not recognized, **what is that large red item to the left ?** The answer of course would be recorded via the **learning object**, along with the newly saved image and accompanying information. Obviously, the same scenario would occur if there was an unrecognized smell or sound.

Just to expand a little more on this particular scenario, if a human said the big red item to the left in the scene was a **truck**. The word **truck** would be looked up online and the definition and an image of it would be saved with the current scene information. The analysis object would send a message to the audio object to say **thank you for the clarification**.

If the definition and image could not be found, the analysis object would send a message to the audio object to for clarification. For example, the Android could say *couldn't find the meaning of a widget online, please spell it for me*, or *couldn't find the meaning of a widget, could it be something else ?*

If an unknown item is found in the current scene and there is no one to ask, the analysis object will send a request to the online service to try to match as close as possible the image of the unknown item. This is where those **artificial neural nets** earn their keep. If successful, the same saving of the information just described above would occur. If the online search is unsuccessful a local image search of the closest match would be done and if something is found, it would be saved as before, but with a **questionable** or **unconfirmed notation** next to it.

Finally, a closer inspection may be required, probably from different angles of view. If the new view yields a different object, the *identification search* as described above would be repeated with the slightly different view of the item. Failing any matching at all, nothing can be done and the issue will be abandoned.

To illustrate the significance of the above described procedures, consider how children learn, by touching, smelling, listening and asking questions: *Why is the sky blue ? or what is that fuzzy thing with four legs and a tail ? why did it bite me ?*

As time goes on, the child physically grows in size and knowledge, but still has questions: *How far is the moon from earth ? Why do I have to go to sleep now ?* As the youngster grows, so do the questions. All the time, asking question, investigating and storing the information. In their teens they realize how dumb their parents are. Usually, by their thirties, they realize how smart their parents are. And so it goes.

By thirty, forty, fifty or sixty the questions of the unknown gets less and less because the database has more and more information stored. Sometimes however, we forget things we used to know, just like the **memory object** which simulates the same effect in this system when it culls the infrequently used memory storage.

Learning occurs when an unfamiliar situation is encountered. We scan our memory banks looking for similar events to make assumptions about the current situation. This is what is intended to be simulated.

b.   Formal Learning

The self learning android, by definition, would have to go to school. Probably home school. In a classroom type environment, with a teacher explaining the subject matter verbally and on the blackboard, the android would have to be able to follow along. Therefore, the **visual** and **audio** objects would be sending information to the **analysis object.**

The analysis object in a similar method as cited before would be comparing the incoming information to previously stored information using the **memory object**. Again, if the information is clearly new, it would be saved. If there is some ambiguity, the Android would ask the Teacher for clarification. The android would have to be able to *read* the blackboard and assigned books in print or online, like a computer or tablet screen.

c.   Decisions, Decisions, Decisions

How do we make decisions? We all know when we are making decisions: Like picking a school for our child to attend or which restaurant to go to for dinner tonight or can I get away without cutting the grass till next weekend? or should I do my homework on a Friday night or go out with my friends ? or should I buy the 50 inch or 60 inch flat screen TV ?

In each case, we define the problem, then usually list the pros and cons to each choice for the resolution of the problem. Based on the relative importance of each pro and each con, we usually go for the least costly in terms of time, effort, financial cost or downside risk. Whichever of the parameters or the combination of some or all of these parameters, is most important to us, determines our choice. (refer to **Critical Thinking** cited in chapter **3** section **f**)

How many decisions do we make? Actually we make tons of "mini" decisions all the time. We make them so often, so quickly and they are so small, we do them by rote and we don't even know we're doing it. Some examples would be: should I hold the door for the person walking behind me? Should I step on the gas to get past that amber traffic light? Paper or plastic? Should I take a sip of coffee now or wait ? (the android doesn't drink coffee) Should I say good morning to people at work ?

As with the big decisions or small decisions, the process is the same. We define the problem, we list the pros and cons for each possible solution. We then assign relative values on the resulting parameters as before: estimating the cost in time, effort, financial cost and downside risk.

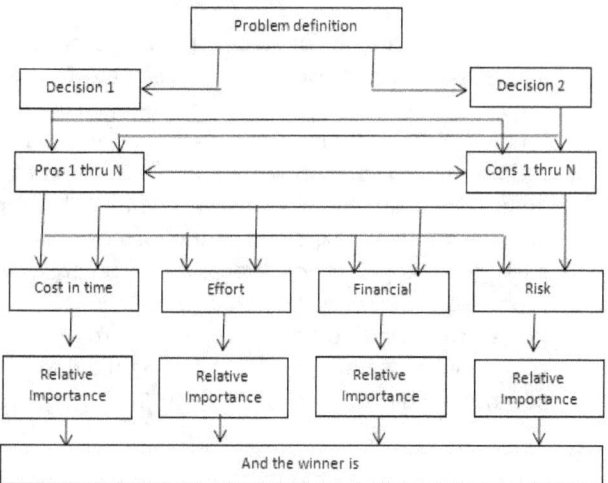

Figure 38 - Critical Thinking Flow Diagram

Several analysis objects are needed at each step in this process along with a database entry for the given scenario. for example, the **holding the door, the traffic light, the sip of coffee** scenarios.

➢ The **first step, define the problem.** In order for the analysis object to define the problem the current environment or situation must be known. Therefore the visual, audio, odor and tactile objects are part of this procedure. Examples of the problem could be: is the Android walking and opening a door and someone is behind them ? Is the android driving a car (I hope they have a driver's license) ? Is the android at work?

➢ **Step two** is define the **possible solutions.** At least two or more solutions to the problem are usually possible. If there is only one solution, there is no further processing required.

➢ **Step three, for each possible solution define the pros and cons.** There may be several of each. The analysis object(s) must define the pros and cons. If there are only pros or only cons, the decision is complete

➤ The **fourth step** is for each pro and each con, the analysis object(s) must assign an estimate of the **time** required, the **effort** required, the financial **cost** and the **relative risk** of failure. Obviously, the analysis object estimates each category separately. This is one of those items which are honed with experience. This type of information would be saved in the system database for each pro and con. For example, the analysis object estimates the door would need to be held two seconds longer, the effort to hold the door is zero, no financial cost is involved and the risk for holding the door is none but not holding the door is high. This is where the experience comes in.

➤ The next **step, number five** is assigning the **relative importance** to each of the estimated parameters. This again is saved in the database specifically for this case, i.e. the **holding the door scenario**. As experience is gained the importance of these items in particular situations comes into play.

➤ Finally the decision is made and the resulting action is taken, by sending a command message to the appropriate object(s).

Note: it is envisioned that multiple analysis objects are working simultaneously in steps 2, 3, 4 and 5, cited above. This parallel processing is for speed of decisions. Further, note in step 4: **time, effort, financial cost** and **risk** may not be the only criteria to evaluate the best decision on, but are probably the minimum.

Just as children may not make the best decisions, but improve their decision making prowess the older they get and the more experience they accumulate, so too would a self learning system improve with experience. Obviously, the important aspect is to have a generic process or mechanism in place to handle **decision making** and which experience can improve the quantity and quality of the decisions as time goes on.

Recall that the visual system automatically records images when the current scene has changed by 5 percent (refer to chapter 8a). That image along with various other data is also sent to the critical thinking process for evaluation.

The critical thinking process would be widely used throughout the system. Certainly during conversations, or classroom teaching sessions.

d. The data in the database

What may offer some further clarification to the operation of the critical thinking process is what is saved in the database. Aside from the obvious information like images from the video processor, the image name, time date of the image, classification of the image location (indoor, outdoor, desert), sound, smell. A list of all known physical objects visible within the image should be include along with links to the characteristic of each of the physical objects in the list (animal, mineral, vegetable, etc.). Further information would be the definition of the physical object, typical size, where found, edible, difficulty rating, critical thinking parameters (time, effort, cost, risk) and probably much more.

Some images in the database will be simple entries, like a picture of an apple, retrieved from the internet when the system learns the single word apple and its meaning and characteristics. Some entries in the database maybe just an odor or a sound, along with their identification, meaning and characteristics.

Chapter 17 … Other Issues

This **modeling** of an android provides a functional description of the elements or mechanisms which attempts to facilitate mimicking human behavior. In that attempt, the systems design allows for self learning to take place. The more traditional method of designing a computerized android would seem a much more daunting task. Since the design would have to cover any and all variations in an uncountable number of tasks.

The *self learning* approach is one level down from the traditional design method of simply mimicking the observable human actions. The self learning approach attempts simulate in theory, the underlying processes or mechanisms which allow the observable higher level behavior to occur, in other words, the embedded rules which the behavior follows.

Self learning is the key, even with all the information in the world, nothing would happen unless the system had motivations or goals and had the ability to do something (carry out actions) with the information. Of course, with the self learning approach, the Android would need to be taught and learn via it's own experiences.

This system provides a *virtually continuous learning mode of operation*. Learning takes place on a physical (motion), visual , verbal, odor, sound and touch basis. In other words, the Android learns by doing, seeing, discussing (including a classroom environment), smelling and hearing, very much like the androids human counterpart.

This system model further provides for safeguards in the form of a high level overarching goal of not harming any living thing and always trying to not dis-please the human. Another generally restrictive factor may be having only a critical thinking algorithm in-place. The critical thinking algorithm chooses between two or more previously learned or stored paths. Whereas, creative, intuitive or conceptual thinking develops new ideas based on combinations of previously learn items and is much less restrictive.

The following is from Bill L'Hotta 06/2019 - former Bell Labs engineer:

> Some areas where I see predictive "thinking" in computers already:
>
> My garmin watch knows where I go on Friday mornings and tells me how long my commute is
> Web browsers pre-populate suggested web sites based on your history (if you turn on that option)
> Email servers filter out spam via smart algorithms.

Considering the current state of the art regarding things like a Facial Recognition, GPS, Echo Dot[8], Google Home[9], etc. and driver-less cars, we are on our way to the autonomous android. The only question remains, where to get the power for all those *muscles* in the android's body. Tony Stark[10] used a nuclear power supply, perhaps our Android, Bob could as well.

---

[8] Amazon' Alexa
[9] Google's version of Alexa
[10] The Iron Man fictional character

Table of figures

Figure 1   - Human Brain Early Development..................3
Figure 2   - Human Brain vs Animal Brain.......................4
Figure 3   - Comparison of Neuron in Cerebral Cortex....6
Figure 4   - Motor with tachometer feedback..............28
Figure 5   - Letter A - 5x7 matrix...................................30
Figure 6   - Letter B - 5x7 matrix...................................31
Figure 7   - Letter C - 5x7 matrix...................................31
Figure 8   - Simple Neural Network Diagram.................31
Figure 9   - Samples Letter A...32
Figure 10   - Neural Network Learning Diagram............32
Figure 11   - Fuzzy Logic..35
Figure 12   - Boolean Logic Diagram...............................35
Figure 13   - Fuzzy Logic Diagram.................................. 36
Figure 14   - Companies using Fuzzy Logic............36
Figure 15   - Companies using Fuzzy Logic....................36
Figure 16   - Binary Search Database Example..............38
Figure 17   - Depth First Search Example.......................39
Figure 18   - Breadth First Search Diagram....................40
Figure 19   - Heuristics Search Diagram.........................41
Figure 20   - Start position and end goal position of tiles42
Figure 21   - Step one of search......................................43
Figure 22   - Results of step 1 of search.........................43
Figure 23   - Step 2 of search.. 44
Figure 24   - Step 3 of search.. 45
Figure 25   - Results of step 3 of search.........................45
Figure 26   - Start and End State......................................48
Figure 27   - Results after step 1 of heuristic search.....49
Figure 28   - Results after step 2 of heuristic search.....49
Figure 29   - Results after step 3 of heuristic search.....50
Figure 30   - Neural Network with hidden layer............52
Figure 31   - Goal System Diagram................................. 65
Figure 32   - Self Learning Cell Diagram.........................68
Figure 33   - Block Diagram of Self Learning Object...... 70

Figure 34   - Communications between objects when
        walking...75
Figure 35   - Hand and finger motion objects................77
Figure 36   - Arm to wrist motion objects..................... 78
Figure 37   - Playing baseball objects........................... 92
Figure 38   - Critical Thinking Flow Diagram................ 102

References

Cover Photo from Interloveupted.blogspot.com

History of robotics from slideshare.net

Artificial Intelligence - From Wikipedia, the free encyclopedia

Common types of artificial intelligence - Simplicable.com

Recursive self-improvement - posted by John Spacey, March 30, 2016 updated on January 08, 2017

Stages of brain development in an infant - from How it Works Magazine 5/2019 based on a US Government study

Human brain size compared to animals - internet article by By Peter Aldhous 10 Feb 2015

How Many Tasks Can Our Brains Process At The Same Time? - Neuroscientist Harris Georgiou from the National Kapodistrian University of Athens in Greece George Dvorsky 11/10/14 12:30pm Filed to: NEUROSCIENCE

Neuromorphic Chips: Microchips that Imitate the Brain - By UNIVERSITY OF ZURICH JULY 23, 2013

Neuromorphic engineering - Wikipedia, the free encyclopedia

Can Computers Think? by Frank Buytendijk Dec. 27, 2012

The Difference Between Brains and Computers - www.computersciencedegreehub.com

Can Computers Think Creatively? By Mark McGuiness

Definition of thinking and thought - Merriam-Webster Dictionary

Types of thinking - Encyclopedia Britannica

Expert Systems - From Wikipedia, the free encyclopedia

Seven steps in problem solving - identified by Robert J. Sternberg

Applications using Fuzzy logic - Guru99.com

Heuristic technique - From Wikipedia, the free encyclopedia

Solving the tile problem using A* and heuristics by Ajinkya Sonawane - Sept. 15, 2018

Artificial neural networks - Wikipedia, the free encyclopedia

Types of artificial neural networks - Wikipedia, the free encyclopedia

Algorithms used in the artificial neural networks - from Wikipedia, the free encyclopedia

Object Oriented Software - Excepts from www.tutorialspoint.com

Self-management (computer science) - Wikipedia, the free encyclopedia

Goal-Oriented Approach to Self-Managing Systems Design - Steven J. Bleistein and Pradeep Ray. School of Information Systems, Technology, and Management, University of New South Wales, Kensington 2052, NSW, Australia

A Goal-Oriented Approach for the Generation and Evaluation of Alternative Architectures - Gemma Grau, Xavier Franch - Universitat Politècnica de Catalunya (UPC) c/ Jordi Girona 1-3, Barcelona E-08034, Spain.

Some areas where I see predictive "thinking" in computers already - Bill L'Hotta 06/2019 - former Bell Labs engineer

Robotics is about us. It is the discipline of emulating our lives, of wondering how we work. - Roderic Grupen , a University of Massachusetts professor who is heading the robotic learning project.